# REA's
# MATH
# TUTOR®
## for the SAT I

**Staff of Research and Education Association**
**Dr. M. Fogiel, Chief Editor**

*Research and Education Association*
61 Ethel Road West
Piscataway, New Jersey 08854

# REA's MATH TUTOR® for the SAT I

1997 PRINTING

Printed in the United States of America

Library of Congress Catalog Card Number 94-67717

International Standard Book Number 0-87891-962-7

THE HIGH SCHOOL TUTOR is a registered trademark of Research & Education Association, Piscataway, New Jersey 08854

# CONTENTS

# About Research & Education Association

Research and Education Association (REA) is an organization of educators, scientists, and engineers specializing in various academic fields. Founded in 1959 with the purpose of disseminating the most recently developed scientific information to groups in industry, government, and universities, REA has since become a successful and highly respected publisher of study aids, test preps, handbooks, and reference works.

Created to extensively prepare students and professionals with the information they need, REA's Test Preparation series includes study guides for the Tests of General Educational Development (GED), the Scholastic Assessment Tests (SAT), the Advanced Placement Exams (AP), the Test of English as a Foreign Language (TOEFL), as well as the Graduate Record Examinations (GRE), the Graduate Management Admission Test (GMAT), the Law School Admission Test (LSAT), and the Medical College Admission Test (MCAT).

Whereas most Test Preparation books present few practice exams which bear little resemblance to the actual exams, REA's series presents exams which accurately depict the official tests in both degree of difficulty and types of questions. REA's practice tests are always based on the most recently administered tests and include every type of question that can be expected on the actual tests.

REA's publications and educational materials are highly regarded and continually receive an unprecedented amount of praise from professionals, instructors, librarians, parents, and students. Our authors are as diverse as the subjects and fields represented in the books we publish. They are well known in their respective fields and serve on the faculties of prestigious universities throughout the United States.

# Acknowledgments

We would like to thank the following people for their contributions:

Dr. Max Fogiel, President, for his overall guidance which has brought this publication to its completion

Stacey A. Sporer, Managing Editor, for directing the editorial staff throughout each phase of the project

Craig D. Thomason, Project Editor, for coordinating the development of the book

In addition, special thanks go to:

Suzanne Coffield, M.A., Anita Price Davis, Ed.D., Joseph D. Fili, M.A.T., Marilyn B. Gilbert, M.A., Bernice E. Goldberg, Ph.D., Gary Lemco, Ph.D., Richard C. Schmidt, Ph.D.

Clair Adas and Jennifer LoBalbo for their editorial contributions

Marty Perzan for typesetting the book.

# How to Use
# This Book

# HOW TO USE THIS BOOK

## WHAT THIS BOOK IS FOR

For as long as the SAT I has been taken by high school students, they have found this test to be difficult and challenging. Despite the publication of hundreds of test preparation books intended to provide improvement over previous guides, students continue to remain perplexed.

In a study of the problem, REA found the following basic reason underlying students' difficulties with taking the SAT I.

Students need systematic rules of analysis which may follow on a step-by-step manner to solve the usual questions and problems encountered.

This book is intended to aid students in arithmetic, algebra, geometry, and word problems which are found in the SAT I math sections, by supplying detailed explanations which may not be apparent to students. In using this book, students may review and study the illustrated questions at their own pace; they are not limited to the time allowed for explaining questions in a classroom setting.

To meet the objectives of this book, we have selected question types usually encountered on the SAT I examination, and have solved each one meticulously to illustrate the steps which students find difficult to comprehend at times.

### To Learn and Understand a Question Type Thoroughly

This book is set up so that each type of SAT I math question is covered thoroughly. Chapter 1 highlights Regular Math, which allows students to review the concepts that are needed to succeed on these questions. Chapter 2 covers Quantitative Comparisons. This chapter will test your ability to solve problems involving quantities and comparing one mathematical expression to another. Chapter 3 highlights Student-Produced Response questions, in which students enter their answer in a grid, instead of the normal multiple-choice format.

## ABOUT THE SAT I

### Who Takes the Test and What is it Used For?

The SAT I is usually taken by high school juniors and seniors. College admissions officers use the test as a way to fairly judge all the students that apply to their school. Because high schools often have many different grading systems, the officers use SAT I scores to put applicants on equal ground. Your SAT I score, along with other information provided by you and your high school, help colleges predict how well you will do at the college level.

The SAT I is usually a requirement for entering college, but if you do poorly, it does not automatically mean you cannot get into college or that you will not do well once you are there. A score on the SAT I that does not match your expectations does not mean you should change your plans about attending college. There are several other criteria by which admissions officers judge applicants, such as grade point average, extracurricular activities, and course levels taken in high school.

### Who Administers the Test?

The SAT I is developed and administered by the Educational Testing Service (ETS) and involves the assistance of educators throughout the country. The test development process is designed and implemented to ensure that the content and difficulty level of the test are appropriate.

### When Should the SAT I Be Taken?

You should try to take the test early in your junior or senior year so that you will have another opportunity to take it if you are not satisfied with your performance.

### When and Where is the Test Given?

The SAT I is administered seven times a year in most states. It is given at hundreds of locations throughout the country, including high schools. The usual testing day is Saturday, but the test may be taken on an alternate day if a conflict exists, such as a religious obligation.

For information on upcoming administrations of the SAT I, consult the *SAT I Registration Bulletin,* which may be obtained from your guidance counselor or by contacting:

> Educational Testing Service
> P.O. Box 6200
> Princeton, NJ 08541

## Is There a Registration Fee?

To take the SAT I, you must pay a registration fee. A fee waiver may be granted in certain situations. To find out if you qualify, and to register for the waiver, contact your guidance counselor.

## FORMAT OF THE SAT I

The following chart summarizes the math format of the SAT I.

| Section | Question Type | Number of Questions |
|---------|---------------|---------------------|
| **Mathematics** (Skills covered: Arithmetic, Algebra, and Geometry) | Regular Math Quantitative Comparison Critical Reading Student-Produced Response (Grid-Ins) | 35 multiple-choice 15 multiple-choice 10 non-multiple-choice |

There are three types of math questions on the SAT I:

• Regular Math: (35 questions) These are mathematics questions of the standard multiple-choice format.

• Quantitative Comparisons: (15 questions) You will be asked to compare the quantities in column A and B in terms of the four statements presented.

• Student-Produced Response: (10 questions) These questions require you to solve a problem ard enter the solution into a grid. You will not be asked to choose a correct answer from among answer choices.

## CALCULATOR USE

The use of calculators is permitted during the test. You may use a programmable or non-programmable four-function, scientific, or graphing calculator. No pocket organizers, hand-held minicomputers, paper tape, or noisy calculators may be used. In addition, calculators requiring an external power source will not be permitted. You will not be allowed to share a calculator, so bring your own.

# Chapter 1
# Regular Math

# CHAPTER 1

# REGULAR MATH

The Regular Math questions of the SAT I are designed to test your ability to solve problems involving arithmetic, algebra, and geometry. A few of the problems may be similar to those found in a math textbook and will require nothing more than the use of basic rules and formulas. Most of the problems, however, will require more than that. Regular Math questions will ask you to think creatively and apply basic skills to solve problems.

All Regular Math questions are in a multiple-choice format with five possible responses. There are a number of advantages and disadvantages associated with multiple-choice math tests. Learning what some of these advantages and disadvantages are can help you improve your test performance.

The greatest disadvantage of a multiple-choice math test is that every question presents you with four wrong answers. These wrong answers are not randomly chosen numbers—they are the answers that students are most likely to get if they make certain mistakes. They also tend to be answers that "look right" to someone who does not know how to solve the problem. Thus, on a particular problem, you may be relieved to find "your" answer among the answer choices, only to discover later that you fell into a common error trap. Wrong answer choices can also distract or confuse you when you are attempting to solve a problem correctly, causing you to question your answer even though it is right.

The greatest *advantage* of a multiple-choice math test is that the *right* answer is also presented to you. This means that you may be able to spot the right answer even if you do not understand a problem completely or do not have time to finish it. It means that you may be able to pick the right answer by guessing intelligently. It also means that you may be saved from getting a problem wrong when the answer you obtain is not among the answer choices—and you have to go back and work the problem again.

Keep in mind, also, that the use of a calculator is permitted during the test. Do not be tempted, however, to use this as a crutch. Some problems can actually be solved more quickly without a calculator, and you still have to work through the problem to know what numbers to punch. No calculator in the world can solve a problem for you.

## ABOUT THE DIRECTIONS

The directions found at the beginning of each Regular Math section are simple—solve each problem, then mark the best of five answer choices on your answer sheet. Following these instructions, however, is important information that you should understand thoroughly before you attempt to take a test. This information includes definitions of standard symbols and formulas that you may need in order to solve Regular Math problems. The formulas are given so that you don't have to memorize them—however, in order to benefit from this information, you need to know what is and what is not included. Otherwise, you may waste time looking for a formula that is not listed, or you may fail to look for a formula that is listed. The formulas given to you at the beginning of a Regular Math section include:

- The number of degrees in a straight line

- Area and circumference of a circle; number of degrees in a circle

- Area of a triangle; Pythagorean Theorem for a right triangle; sum of angle measures of a triangle

Following the formulas and definitions of symbols is a very important statement about the diagrams, or figures, that may accompany Regular Math questions. This statement tells you that, unless stated otherwise in a specific question, figures are drawn to scale.

## ABOUT THE QUESTIONS

Most Regular Math questions on the SAT I fall into one of three categories: arithmetic, algebra, and geometry. In the following three sections, we will review the kinds of questions you will encounter on the actual test.

# ARITHMETIC QUESTIONS

Most arithmetic questions on the SAT I fall into one of the following four question types. For each question type, an example and solution will be given, highlighting strategies and techniques for completing the problems as quickly as possible.

## Question Type 1: Evaluating Expressions

Arithmetic questions on the SAT I often ask you to find the value of an arithmetic expression or to find the value of a missing term in an expression. The temptation when you see one of these expressions is to calculate its value—a process that is time-consuming and can easily lead to an error. A better way to approach an arithmetic expression is to use your knowledge of properties of numbers to spot shortcuts.

### PROBLEM

$7(8 + 4) - (3 \times 12) =$

(A)  24

(B)  48

(C)  110

(D)  144

(E)  3,024

### SOLUTION

Before you jump into multiplication, look at the numbers inside the first parentheses. This is the sum $(8 + 4) = 12$, which makes the entire expression equal to $7(12) - (3 \times 12)$. The distributive property tells you that $a(b + c) = ab + ac$ and $a(b - c) = ab - ac$. The expression $7(12) - (3 \times 12)$ can be made to fit the second formula, with $a$ equal to 12 and 7 and 3 equal to $b$ and $c$, respectively. Thus, $7(12) - (3 \times 12)$ becomes $12(7 - 3)$, and the answer is simply $12 \times 4$, or 48.

## Question Type 2: Undefined Symbols

Most SAT I math sections include problems that involve undefined symbols. In some problems, these symbols define a value by asking you to perform several arithmetic operations. For example, the symbol $\boxed{x}$ may tell you to square some number $x$ then subtract 3: $\boxed{x} = x^2 - 3$. In other problems, a symbol may represent a missing numeral, such as $10 - \Delta = 7$. By looking at the arithmetic, you can see that $\Delta$ must equal 3.

**PROBLEM**

Let $[n] = n^2 + 1$ for all numbers $n$. Which of the following is equal to the product of $[2]$ and $[3]$?

(A) $[6]$            (D) $[9]$

(B) $[7]$            (E) $[11]$

(C) $[8]$

**SOLUTION**

The newly defined symbol is $[\ ]$. To find the values for $[2]$ and $[3]$, plug them into the formula $[n] = n^2 + 1$.

$$[2] = 2^2 + 1 = 4 + 1 = 5$$

$$[3] = 3^2 + 1 = 9 + 1 = 10$$

Since $[2] = 5$, and $[3] = 10$, we can compute the product: $5 \times 10 = 50$.

Now look at the answers. You will see that the answers are given in terms of $[\ ]$. Once again, you must plug them into the formula $[n] = n^2 + 1$. If we plug each answer choice into the equation, we get

$$[6] = 6^2 + 1 = 36 + 1 = 37$$

$$[7] = 7^2 + 1 = 49 + 1 = 50$$

$$[8] = 8^2 + 1 = 64 + 1 = 65$$

$$[9] = 9^2 + 1 = 81 + 1 = 82$$

$$[11] = 11^2 + 1 = 121 + 1 = 122$$

## Question Type 3: Averages

Some SAT I math problems ask you to simply compute the average of a given set of values. More challenging problems ask you to apply the definition of average. You will recall that the average of a given set of values is equal to the sum of the values divided by the number of values in the set.

**PROBLEM**

The average of 10 numbers is 53. What is the sum of the numbers?

(A)  106

(B)  350

(C)  363

(D)  530

(E)  615

**SOLUTION**

You can solve this problem using the formula

$$\text{Average} = \frac{\text{sum of values}}{\text{number of values}}.$$

In this question,

$$53 = \frac{\text{sum of values}}{10}.$$

Therefore, the sum of the numbers is 530.

$$10 \times 53 = \text{sum of values}$$

$$530 = \text{sum of values}$$

**PROBLEM**

The average of three numbers is 16. If one of the numbers is 5, what is the sum of the other two?

(A)  11

(B)  24

(C)  27

(D)  38

(E)  43

**SOLUTION**

In this problem, if we plug in the numbers we know, we get

$$16 = \frac{a + b + c}{3}.$$

This can be converted to

$$3 \times 16 = a + b + c,$$

which means that

$$48 = a + b + c.$$

If $a = 5$, we can solve:

$$48 = 5 + b + c, \text{ or } 48 - 5 = b + c,$$

and finally $43 = b + c$. Therefore, the sum of the other two numbers is 43 which is choice (E). Notice that choice (A) is waiting for the person who fails to work the formula and simply subtracts 5 from 16.

## Question Type 4: Data Interpretation

Data interpretation problems usually require two basic steps. First, you have to read a chart or graph in order to obtain certain information. Then you have to apply or manipulate the information in order to obtain an answer.

*PROBLEM*

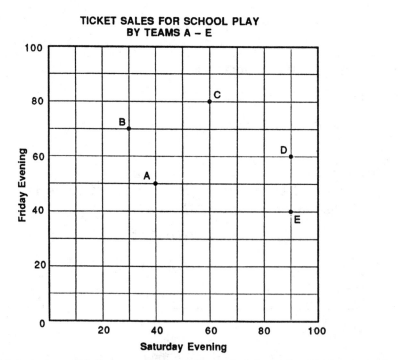

TICKET SALES FOR SCHOOL PLAY
BY TEAMS A – E

Which team sold the greatest number of tickets for Friday evening and Saturday evening combined?

(A)  Team A

(B)  Team B

(C)  Team C

(D)  Team D

(E)  Team E

### SOLUTION

Glancing over the data, you see that the number of tickets sold for Friday evening is represented vertically, while the number of tickets sold for Saturday evening is represented horizontally. Points placed on the grid represent each team's ticket sales for the two evenings.

Read the graph to determine the number of tickets sold by each team for Friday evening and Saturday evening.

Add each pair of numbers to find the total number of tickets sold by each team for both evenings.

Compare the totals to see which team sold the most tickets.

The answer is (D), since team D sold 60 tickets for Friday evening and 90 tickets for Saturday evening for a highest total of 150.

# ALGEBRA QUESTIONS

Algebra problems use letters or variables to represent numbers. In these types of problems, you will be required to solve existing algebraic expressions or translate word problems into algebraic expressions.

## Question Type 1: Algebraic Expressions

Problems involving algebraic expressions often contain hidden shortcuts. You can find these shortcuts by asking yourself, "How can this expression be rearranged?" Often rearrangement will cause an answer to appear almost magically.

There are three basic ways in which you can rearrange an algebraic expression. You can

1. combine like terms.

2. factor the expression.

3. multiply out the expression.

### PROBLEM

If $x = \frac{1}{2}$, which of the following equals $x^2 - x + \frac{1}{4}$?

(A) $-\dfrac{1}{2}$   (D) $\dfrac{1}{2}$

(B) 0   (E) 1

(C) $\dfrac{1}{4}$

### SOLUTION

You can find the answer by substituting $x = \frac{1}{2}$ into the given expression, but there is an easier way. Look at $x^2 - x + \frac{1}{4}$. Remembering the strategy tip, this expression is equal to the trinomial square $(x - \frac{1}{2})^2$. Now you can see at a glance that since you are told that

$$x = \frac{1}{2}, (x - \frac{1}{2})^2$$

must equal 0. Thus, the answer is (B).

## Question Type 2: Word Problems

Among the most common types of word problems found on the SAT I are age problems, mixture problems, distance problems, and percent problems. You can find detailed explanations of these and other problem types in the Basic Math Skills Review. There is a strategy, however, that can help you to solve all types of word problems—learning to recognize "keywords."

Keywords are words or phrases that can be translated directly into a mathematical symbol, expression, or operation. As you know, you usually cannot solve a word problem without writing some kind of equation. Learning to spot keywords will enable you to write the equations you need more easily.

Listed below are some of the most common keywords. As you practice solving word problems, you will probably find others.

| Keyword | Mathematical Equivalent |
|---|---|
| is | equals |
| sum | add |
| plus | add |
| more than, older than | add |
| difference | subtract |
| less than, younger than | subtract |
| twice, double | multiply by 2 |
| half as many | divided by 2 |
| increase by 3 | add 3 |
| decrease by 3 | subtract 3 |

**PROBLEM**

Adam has 50 more than twice the number of "frequent flier" miles that Erica has. If Adam has 200 frequent flier miles, how many does Erica have?

(A) 25                                   (D) 100

(B) 60                                   (E) 250

(C) 75

**SOLUTION**

The keywords in this problem are "more" and "twice." If you let $a =$ the number of frequent flier miles that Adam has and $e =$ the number of frequent flier miles that Erica has, you can write: $a = 50 + 2e$. Since $a = 200$, the solution becomes: $200 = 50 + 2e$. Therefore, $200 - 50 = 2e$, or $150 = 2e$, and $^{150}/_2 = e$ or $75 = e$, which is choice (C).

## GEOMETRY QUESTIONS

SAT I geometry questions require you to find the area or missing sides of figures given certain information. These problems require you to use "if . . . then" reasoning or to draw figures based on given information.

### Question Type 1: "If . . . Then" Reasoning

You will not have to work with geometric proofs on the SAT I, but the logic used in proofs can help you enormously when it comes to solving SAT I geometry problems. This type of logic is often referred to as "If . . . then" reasoning. In "if . . . then" reasoning, you say to yourself, "If A is true, then B must be true." By using "if . . . then" reasoning, you can draw conclusions based on the rules and definitions that you know. For example, you might say, "If ABC is a triangle, then the sum of its angles must equal 180°."

**PROBLEM**

If triangle *QRS* is an equilateral triangle, what is the value of *a* + *b*? (See figure on the following page.)

(A) 60°                                  (D) 100°

(B) 80°                                  (E) 120°

(C) 85°

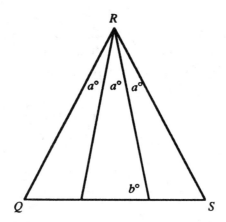

## SOLUTION

You can obtain the information that you need to solve this problem by using a series of "if . . . then" statements: "If *QRS* is an equilateral triangle, then each angle must equal 60°." "If angle *R* = 60°, then *a* must equal 20°." "If *a* = 20°, then *b* must equal half of (180° − 20°) or 80°." "If *a* = 20° and *b* = 80°, then *a* + *b* = 100°." Therefore, answer (D) is correct.

## Question Type 2: Drawing Diagrams

Among the most difficult geometry problems on the SAT I are those that describe a geometric situation without providing a diagram. For these problems, you must learn to draw your own diagram based on the information that is given. The best way to do this is step-by-step, using each piece of information that the problem provides. As you draw, you should always remember to:

1. label all points, angles, and line segments according to the information provided.

2. indicate parallel or perpendicular lines.

3. write in any measures that you are given.

## PROBLEM

If vertical line segment $\overline{AB}$ is perpendicular to line segment $\overline{CD}$ at point *O* and if ray *OE* bisects angle *BOD,* what is the value of angle *AOE*?

(A)  45°

(B)  90°

(C)  120°

(D)  135°

(E)  180°

## SOLUTION

Draw as follows:

Draw and label vertical line segment $\overline{AB}$.

Draw and label line segment $\overline{CD}$ perpendicular to $\overline{AB}$. Label the right angle that is formed. Label point $O$.

Locate angle $BOD$. Draw and label ray $OE$ so that it bisects, or cuts into two equal parts, angle $BOD$. Use equal marks to show that the two parts of the angle are equal. Since you are bisecting a right angle, you can write in the measure $45°$.

Your diagram should resemble that shown below. Now you can evaluate your drawing to answer the question. Angle $AOE$ is equal to $90° + 45°$, or $135°$. The answer is (D).

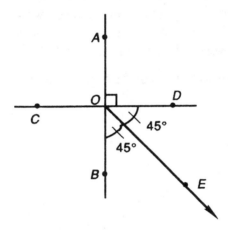

# ANSWERING REGULAR MATH QUESTIONS

The following steps should be used to help guide you through answering Regular Math questions. Combined with the review material which you have just studied, these steps will provide you with the tools necessary to correctly answer the questions you will encounter.

STEP 1   Try to determine the type of question with which you are dealing. This will help you focus in on how to attack the question.

STEP 2   Carefully read all of the information presented. Make sure you are answering the question, and not incorrectly reading the question. Look for key-words that can help you determine what the question is asking.

| STEP 3 | Perform the operations indicated, but be sure you are taking the easiest approach. Simplify all expressions and equations before performing your calculations. Draw your own figures if a question refers to them, but does not provide them. |

| STEP 4 | Try to work backwards from the answer choices if you are having difficulty determining an answer. |

| STEP 5 | If you are still having difficulty determining an answer, use the process of elimination. If you can eliminate at least two choices, you will greatly increase your chances of correctly answering the question. Eliminating three choices means that you have a fifty-fifty chance of correctly answering the question if you guess. |

| STEP 6 | Once you have chosen an answer, fill in the oval which corresponds to the question and answer which you have chosen. Beware of stray lines on your answer sheet, as they may cause your answers to be scored incorrectly. |

Now, use the information you have just learned to answer the following questions.

**DIRECTIONS:** Solve each problem, using any available space on the page for scratch work. Then decide which answer choice is the best and fill in the corresponding oval on the answer sheet.

**NOTES:**

(1) The use of a calculator is permitted. All numbers used are real numbers.

(2) Figures that accompany problems in this test are intended to provide information useful in solving the problems. They are drawn as accurately as possible EXCEPT when it is stated in a specific problem that the figure is not drawn to scale. All figures lie in a plane unless otherwise indicated.

**REFERENCE INFORMATION:**

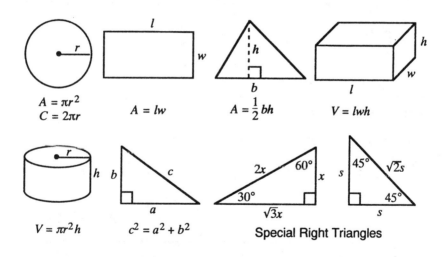

$A = \pi r^2$
$C = 2\pi r$

$A = lw$

$A = \frac{1}{2}bh$

$V = lwh$

$V = \pi r^2 h$

$c^2 = a^2 + b^2$

Special Right Triangles

The number of degrees of arc in a circle is 360.
The measure in degrees of a straight angle is 180.
The sum of the measures in degrees of the angles of a triangle is 180.

## • PROBLEM 1-1

If $f(x) = 3x^2 - x + 5$, then $f(3) =$

(A)  15.         (D)  27.

(B)  17.         (E)  29.

(C)  23.

## SOLUTION:

**(E)**  You are given the function $f(x) = 3x^2 - x + 5$.

To find $f(3)$ means to find out what the given function equals when you let $x$ equal 3. Thus, $f(3)$ is found by substituting 3 in place of $x$, everywhere that $x$ appears

$$f(3) = 3x^2 - x + 5$$
$$f(3) = 3 \times 3^2 - 3 + 5$$
$$= 3 \times 9 - 3 + 5$$
$$= 27 - 3 + 5$$
$$= 24 + 5$$
$$= 29$$

So, the correct answer is (E).

Note:  $f(x)$ is pronounced "$f$ of $x$," and

$f(3)$ is pronounced "$f$ of 3."

## • PROBLEM 1-2

A runner takes 9 seconds to run a distance of 132 feet. What is the runner's speed in miles per hour?

(A)  9         (D)  12

(B)  10         (E)  13

(C)  11

## SOLUTION:

**(B)**  The runner's speed is $\dfrac{132 \text{ feet}}{9 \text{ sec.}}$.

Speed is always a distance (in this case "feet") divided by a time (in this case "seconds"). Note, though, that the problem asks for speed in miles per hour, *not* in feet per second. Thus, the rest of the problem consists of converting to the desired miles per hour.

Here's our plan:

*Step I*—Convert the numerator from feet to miles.

*Step II*—Convert the denominator from seconds to hours.

*Step III*—Divide the new numerator by the new denominator. (Remember that "per" indicates division, so miles per hour means miles/hour).

*Step I*

(a)  Recall that 5,280 feet = 1 mile.

(b)  Multiply (132 feet) by the conversion factor $\left( \dfrac{1 \text{ mile}}{5,280 \text{ feet}} \right)$.

$$(132 \text{ feet}) \times \left( \frac{1 \text{ mile}}{5,280 \text{ feet}} \right) = \frac{132}{5,280} \text{ mile}$$

$$= .025 \text{ mile}$$

(c)  Notice that we cancelled the word "feet" in the numerator with the word "feet" in the denominator, *just like they were numbers.* This left us with "miles" in the numerator, which is what we wanted.

(d)  In step (b) above, we were careful to choose the conversion factor

$$\left( \frac{1 \text{ mile}}{5,280 \text{ feet}} \right)$$

rather than the other possibility

$$\left( \frac{5,280 \text{ feet}}{1 \text{ mile}} \right),$$

so "feet" would cancel.

*Step II*

Next, we convert (9 sec.) to hours.

(a)   Recall that 1 hour = 60 min., and 1 min. = 60 sec.

(b)   Multiply (9 sec.) by two conversion factors.

$$(9 \text{ sec.}) \times \left( \frac{1 \text{ min.}}{60 \text{ sec.}} \right) \times \left( \frac{1 \text{ hour}}{60 \text{ min.}} \right) = \frac{9}{60 \times 60} \text{ hour}$$

$$= \frac{9}{3,600} \text{ hour}$$

$$= \frac{1}{400} \text{ hour}$$

$$= .0025 \text{ hour}$$

*Step III*

Combining the results from Step I and Step II gives

$$\left( \frac{132 \text{ feet}}{9 \text{ sec.}} \right) = \frac{.025 \text{ mile}}{.0025 \text{ hour}} = 10 \text{ miles/hour.}$$

So, the correct answer is (B).

Note: The entire problem can be done in one step, but you should only do this if you feel able to keep track of all the factors.

$$\left( \frac{132 \text{ feet}}{9 \text{ sec.}} \right) \left( \frac{1 \text{ mile}}{5,280 \text{ feet}} \right) \left( \frac{60 \text{ sec.}}{1 \text{ min.}} \right) \left( \frac{60 \text{ min.}}{1 \text{ hour}} \right)$$

$$= \frac{132 \times 60 \times 60}{9 \times 5,280} \text{ miles/hour}$$

$$= 10 \text{ miles/hour}$$

## • PROBLEM 1–3

For the triangle pictured below, the degree measure of the three angles are $x$, $3x$, and $3x + 5$. Find $x$.

(A) 25

(B) 27

(C) 28

(D) 28.3

(E) 29

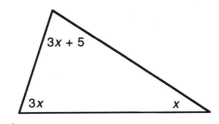

### SOLUTION:

**(A)**     Every triangle has three angles inside.

When you add together the measures of these three angles, you will always get 180°, no matter what size or shape the triangle is.

first angle + second angle + third angle = 180

$$x + 3x + (3x + 5) = 180$$

Group like terms together (terms containing an $x$).

$$(x + 3x + 3x) + 5 = 180$$

Add like terms.

$$7x + 5 = 180$$

Subtract 5 from both sides.

$$7x + 5 - 5 = 180 - 5$$

$$7x = 175$$

Divide both sides by 7.

$$\frac{7x}{7} = \frac{175}{7}$$

$$x = 25$$

So, the correct answer is (A).

## • PROBLEM 1-4

Divide $3\frac{1}{5}$ by $1\frac{1}{3}$.

(A) $2\frac{2}{5}$        (D) $4\frac{4}{15}$

(B) $3\frac{1}{15}$        (E) $8$

(C) $3\frac{3}{5}$

## SOLUTION:

**(A)** We are asked to find $\dfrac{3\frac{1}{5}}{1\frac{1}{3}}$.

In order to do this, we must first change both mixed numbers into improper fractions.

$$3\frac{1}{5} = \frac{(3 \times 5) + 1}{5} = \frac{15 + 1}{5} = \frac{16}{5}$$
$$1\frac{1}{3} = \frac{(1 \times 3) + 1}{3} = \frac{3 + 1}{3} = \frac{4}{3}$$

So, $\dfrac{3\frac{1}{5}}{1\frac{1}{3}} = \dfrac{\frac{16}{5}}{\frac{4}{3}}$.

Observe that $\dfrac{\frac{16}{5}}{\frac{4}{3}}$ is in the form $\dfrac{\frac{a}{b}}{\frac{c}{d}}$.

Also, recall that $\dfrac{\frac{a}{b}}{\frac{c}{d}} = \dfrac{a}{b} \times \dfrac{d}{c}$.

Thus, $\dfrac{\frac{16}{5}}{\frac{4}{3}} = \dfrac{16}{5} \times \dfrac{3}{4}$

$$= \frac{16 \times 3}{5 \times 4}$$

$$= \frac{4 \times 4 \times 3}{5 \times 4}$$

(Wrote 16 as 4 × 4)

$$= \frac{12}{5}$$

(Cancelled 4 from the numerator and denominator)

The final step is to change $^{12}/_5$ (an improper fraction) into a mixed number by dividing the denominator, 5, into the numerator, 12.

$$\frac{12}{5} = 2 + \frac{2}{5} = 2\frac{2}{5}$$

So, the correct answer is (A).

## • PROBLEM 1-5

Change 125.937% to a decimal.

(A) 1.25937          (D) 1,259.37

(B) 12.5937          (E) 12,593.7

(C) 125.937

## SOLUTION:

(A)     To change a percent to a decimal, drop the percent sign and move the decimal point two place values to the left.

You can see why this is true:

$$125.937\% = \frac{125.937}{100}$$

(Definition of percent)

$$= 1.25937$$

(Division by 100 moves decimal point two places to the left)

So, the correct choice is (A).

## • PROBLEM 1–6

Evaluate $4(a + b) + 2[5 - (a^2 + b^2)]$, if $a = 2$ and $b = 1$.

(A) 6　　　　　　　　　　　(D) 20

(B) 7　　　　　　　　　　　(E) 62

(C) 12

### SOLUTION:

**(C)**　　Every place you see an $a$, substitute 2.

Every place you see a $b$, substitute 1.

Remember to:

Simplify quantities inside parentheses first.

Simplify inner parentheses before outer parentheses.

$$4 (a + b) + 2 [5 - (a^2 + b^2)] \quad \text{(Given)}$$

$$= 4 (2 + 1) + 2 [5 - (2^2 + 1^2)] \quad \text{(Substituted for } a \text{ and } b\text{)}$$

$$= 4 (3) + 2 [5 - (4 + 1)]$$

$$= 12 + 2 [5 - (5)]$$

$$= 12 + 2 [0]$$

$$= 12 + 0$$

$$= 12$$

So, the correct choice is (C).

## • PROBLEM 1–7

Find the mean of the following scores:

5, 7, 9, 8, 5, 8, 9, 8, 7, 8, 7, 5, 9, 5, 8, 5, 9, 6, 5

(A) 5　　　　　　　　　　　(D) 8

(B) 6　　　　　　　　　　　(E) 9

(C) 7

## SOLUTION:

**(C)**    We need to apply the definition of the mean to our problem.

$$\text{Mean} = \frac{\text{sum of the scores}}{\text{how many scores there are}}$$

Sum of the scores (add them)

$$= 5 + 7 + 9 + 8 + 5 + 8 + 9 + 8 + 7 + 8 + 7 + 5 + 9$$

$$+ 5 + 8 + 5 + 9 + 6 + 5$$

$$= 133$$

Number of scores (count them) = 19

Thus, mean $= \dfrac{133}{19} = 7$

The correct answer is (C).

## • PROBLEM 1–8

What percent of 260 is 13?

(A)  .05%                (D)  .5%

(B)  5%                  (E)  20%

(C)  50%

## SOLUTION:

**(B)**    We must translate certain keywords from the given word problem into their mathematical equivalents, as follows:

"of" translates to × (multiplication)

"is" translates to = (equals)

"what percent" translates to $x$ (the unknown in decimal form)

Now we can re-write the question as an equation.

(What percent) (of) 260 (is) 13

$$x \times 260 = 13$$

or,        $260x = 13$

$$x = \frac{13}{260} = .05 = \frac{05}{100} = 5\%$$

So, the answer should be (B).

## • PROBLEM 1–9

How many corners does a cube have?

(A)  4

(B)  6

(C)  8

(D)  12

(E)  24

## SOLUTION:

**(C)**    Referring to the figure, we see that the cube has *eight* corners.

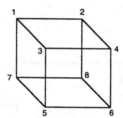

The top has four corners (points 1, 2, 3, and 4).

The bottom has four corners (points 5, 6, 7, and 8).

So, the correct choice is (C).

## • PROBLEM 1–10

Subtract $4\frac{1}{3} - 1\frac{5}{6}$.

(A)  $3\frac{2}{3}$

(B)  $2\frac{1}{2}$

(C)  $3\frac{1}{2}$

(D)  $2\frac{1}{6}$

(E)  None of these.

## SOLUTION:

**(B)** Subtract $4\frac{1}{3} - 1\frac{5}{6}$.

Change the given numbers to improper fractions.

$$4\frac{1}{3} - 1\frac{5}{6} = \frac{(4 \times 3) + 1}{3} - \frac{(1 \times 6) + 5}{6}$$

$$= \frac{13}{3} - \frac{11}{6}$$

$$= \left[\left(\frac{13}{3}\right)\left(\frac{2}{2}\right)\right] - \frac{11}{6} \qquad \text{(Common denominator is 6)}$$

$$= \frac{26}{6} - \frac{11}{6}$$

$$= \frac{26 - 11}{6}$$

$$= \frac{15}{6}$$

Change $^{15}/_6$ into a mixed number.

$$\frac{15}{6} = 2 + \frac{3}{6} = 2 + \frac{1}{2} = 2\frac{1}{2}$$

So, the correct answer is (B).

## • PROBLEM 1–11

The rates of a laundry are $6.25 for the first 15 pieces and $0.35 for each additional piece. If the laundry charge is $8.35, how many pieces were laundered?

(A) 5                 (D) 21

(B) 6                 (E) 25

(C) 15

## SOLUTION:

**(D)** The total laundry charge is $8.35.

Subtracting $6.25 (the cost of the first 15 pieces of laundry) gives:

$8.35 – $6.25 = $2.10.

We want to find how many additional pieces (beyond the first 15) were laundered for this remaining $2.10.

We can use a proportion.

$$\frac{\$.35}{1 \text{ piece}} = \frac{\$2.10}{x \text{ pieces}}$$

Cross multiplying gives

$$.35x = 2.10$$

$$x = \frac{2.10}{.35} = 6 \text{ additional pieces}$$

The total number of pieces is

15 (at a cost of $6.25) + 6 (at a cost of $2.10) = 21 pieces.

So, the choice should be (D).

### • PROBLEM 1–12

The enrollment in Eastern High School is 1,050. If the attendance on a certain day was 94%, how many students were absent that day?

(A)  50

(B)  63

(C)  420

(D)  987

(E)  1,044

## SOLUTION:

**(B)**    If 94% attended, then 6% did *not* attend (were absent). (This was found by subtracting 100% – 94% = 6%.)

Number absent = 6% of total enrollment

$$= 6\% \times 1{,}050$$

$$= .06 \times 1{,}050 \text{ (6\% in its decimal form, is .06)}$$

$$= 63$$

So, the correct answer is (B).

### • PROBLEM 1-13

Find the correct solution to the nearest cent: 15% of $8.75.

(A) $1.31                     (D) $13.13

(B) $1.32                     (E) $131.25

(C) $13.12

**SOLUTION:**

**(A)**     15% of $8.75 = .15 × $8.75

= $1.3125

Since we are working with dollars and cents, we must round this number to two decimal places (the nearest penny). To do this, look at the digit to the right of the second decimal place. This digit is 2. 2 is less than 5, so we round $1.3125 down to $1.31. ($1.3125 is closer to $1.31 than it is to $1.32.)

So, the correct answer is (A).

### • PROBLEM 1-14

Simplify $6\sqrt{7} + 4\sqrt{7} - \sqrt{5} + 5\sqrt{7}$.

(A) $10\sqrt{7}$                     (D) $15\sqrt{16}$

(B) $15\sqrt{7} - \sqrt{5}$                     (E) $60$

(C) $15\sqrt{21} - \sqrt{5}$

**SOLUTION:**

**(B)**     $6\sqrt{7} + 4\sqrt{7} - \sqrt{5} + 5\sqrt{7}$

(Given)

$= (6\sqrt{7} + 4\sqrt{7} + 5\sqrt{7}) - \sqrt{5}$

(Grouped together terms containing $\sqrt{7}$)

$= (6 + 4 + 5)\sqrt{7} - \sqrt{5}$

(Factored out $\sqrt{7}$ from parentheses)

$$= 15\sqrt{7} - \sqrt{5}$$

(Added 6 + 4 + 5 to get 15)

So, the correct answer is (B).

### • PROBLEM 1–15

$4\% \times 4\% =$

(A)  0.0016%                     (D)  16%

(B)  0.16%                       (E)  160%

(C)  1.6%

**SOLUTION:**

**(B)**     $4\% \times 4\% = .04 \times .04$          (Wrote % in decimal form)

$= .0016$

This is the answer in decimal form; notice, though, that your choices are given in % form. Thus, we need to change the decimal to a percent by moving the decimal point two places to the right and adding a percent sign.

$.0016 = .16\%$

So, the correct answer is (B).

### • PROBLEM 1–16

If $x - (4x - 8) + 9 + (6x - 8) = 9 - x + 24$, then $x =$

(A)  4.                          (D)  6.

(B)  2.                          (E)  10.

(C)  8.

**SOLUTION:**

**(D)**     The goal is to get $x$ alone on the left side.

$$x - (4x - 8) + 9 + (6x - 8) = 9 - x + 24 \text{ (Given)}$$

Distribute minus sign into the first set of parentheses. Note that this gives a *positive* 8.

$$x - 4x + 8 + 9 + (6x - 8) = 9 - x + 24$$

Remove the second set of parentheses. They serve no purpose in this equation, except to make it look more complicated than it really is.

$$x - 4x + 8 + 9 + 6x - 8 = 9 - x + 24$$

Using addition and subtraction, move all terms containing $x$ to the left side, and all pure numbers to the right side.

$$x - 4x + 6x + x = 9 + 24 - 8 - 9 + 8$$

Performing the indicated additions and subtractions gives.

$$4x = 24.$$

Dividing both sides by 4 gives.

$$x = 6.$$

So, the correct answer is (D).

### • PROBLEM 1–17

$$\frac{2}{3} + \frac{5}{9} =$$

(A) $\dfrac{7}{12}$     (D) $\dfrac{7}{9}$

(B) $\dfrac{11}{9}$     (E) $\dfrac{11}{3}$

(C) $\dfrac{7}{3}$

### SOLUTION:

**(B)**  A common denominator is needed to add fractions.

The least common denominator in this problem is 9, since the smallest number that both 3 and 9 will divide into is 9.

$$\frac{2}{3} + \frac{5}{9} = \left[\left(\frac{2}{3}\right) \times \left(\frac{3}{3}\right)\right] + \frac{5}{9}$$

(Within the brackets, we are converting $^2/_3$ to ninths)

$$= \left[\frac{2 \times 3}{3 \times 3}\right] + \frac{5}{9}$$

$$= \frac{6}{9} + \frac{5}{9}$$

$$= \frac{6 + 5}{9}$$

$$= \frac{11}{9}$$

So, the correct answer is (B).

### • PROBLEM 1-18

Add $\dfrac{3}{6} + \dfrac{2}{6}$.

(A) $\dfrac{1}{12}$        (D) $\dfrac{8}{9}$

(B) $\dfrac{5}{6}$        (E) $\dfrac{9}{8}$

(C) $\dfrac{5}{12}$

## SOLUTION:

**(B)**  The given fractions already have the same denominator, so they are ready to be added.

$$\frac{3}{6} + \frac{2}{6} = \frac{3 + 2}{6}$$

$$= \frac{5}{6}$$

So, the correct answer is (B).

## • PROBLEM 1–19

A plumber used pieces of pipe measuring $4^{1}/_{4}$ feet, $2^{2}/_{3}$ feet, and $3^{1}/_{2}$ feet. If the pieces of pipe were cut from a 15-foot length of pipe, how many feet of pipe remain? Disregard waste.

(A) $4\dfrac{7}{12}$ ft

(D) $10\dfrac{5}{12}$ ft

(B) $5\dfrac{1}{6}$ ft

(E) $11\dfrac{1}{3}$ ft

(C) $9\dfrac{5}{6}$ ft

## SOLUTION:

**(A)** We want to find how many feet of pipe are left after several pieces have been cut off.

feet of pipe remaining = (original length of pipe) − (lengths of cut off pieces)

$$\text{feet of pipe remaining} = (15) - \left(4\frac{1}{4} + 2\frac{2}{3} + 3\frac{1}{2}\right)$$

Let's start by finding the sum

$$\left(4\frac{1}{4} + 2\frac{2}{3} + 3\frac{1}{2}\right).$$

(Later, we will come back and subtract it from 15.)

$$\left(4\frac{1}{4} + 2\frac{2}{3} + 3\frac{1}{2}\right)$$

(Given sum)

$$= \left[\left(4 + \frac{1}{4}\right) + \left(2 + \frac{2}{3}\right) + \left(3 + \frac{1}{2}\right)\right]$$

(Broke up mixed numbers into whole and fractional parts)

$$= (4 + 2 + 3) + \left(\frac{1}{4} + \frac{2}{3} + \frac{1}{2}\right)$$

(Regrouped terms)

$$= (9) + \left( \frac{1}{4} + \frac{2}{3} + \frac{1}{2} \right)$$

(Added terms in first set of parentheses)

$$= 9 + \left[ \left( \frac{1}{4} \times \frac{3}{3} \right) + \left( \frac{2}{3} \times \frac{4}{4} \right) + \left( \frac{1}{2} \times \frac{6}{6} \right) \right]$$

(Changed fractions to common denominator, which is 12)

$$= 9 + \left( \frac{3}{12} + \frac{8}{12} + \frac{6}{12} \right)$$

$$= 9 + \left( \frac{17}{12} \right)$$

(Added fractions)

$$= 9 + \left( \frac{12}{12} + \frac{5}{12} \right)$$

(Rewrote $^{17}/_{12}$)

$$= 9 + \left( 1 + \frac{5}{12} \right)$$

(Replaced $^{12}/_{12}$ by 1)

$$= 10 + \frac{5}{12}$$

$$= 10 \frac{5}{12}$$

Now, substitute $10^5/_{12}$ back into the original equation.

$$\text{Remaining length of pipe} = 15 - 10 \frac{5}{12}$$

$$= 14 \frac{12}{12} - 10 \frac{5}{12}$$

(Borrowed $^{12}/_{12}$ from 15)

$$= 4\frac{7}{12} \text{ feet}$$

(Subtracted 10 from 14, and $^5/_{12}$ from $^{12}/_{12}$)

So, the correct answer is (A).

### • PROBLEM 1-20

Peter has five rulers of 30 cm each and three of 20 cm each. What is the average length of Peter's rulers?

(A) 25                    (D) 26.25

(B) 27                    (E) 27.25

(C) 23

## SOLUTION:

**(D)**    We apply the definition of the average to our problem.

$$\text{Average} = \frac{\text{sum of the lengths of all the rulers}}{\text{how many rulers there are}}$$

$$= \frac{30 + 30 + 30 + 30 + 30 + 20 + 20 + 20}{8}$$

$$= \frac{(30 \times 5) + (20 \times 3)}{8}$$

$$= \frac{150 + 60}{8}$$

$$= \frac{210}{8}$$

$$= 26.25 \text{ cm}$$

So, the correct answer is (D).

## • PROBLEM 1-21

If $2^{(6x-8)} = 16$, then $x =$

(A)  2.

(D)  1.

(B)  4.

(E)  6.

(C)  10.

## SOLUTION:

**(A)**    First, note that $16 = 2^4$.

(Verify this for yourself: $2^4 = 2 \times 2 \times 2 \times 2 = 16$.)

Next, substitute $2^4$ for 16 in the given equation.

$$2^{(6x-8)} = 16 \qquad \text{(Given)}$$

$$2^{(6x-8)} = 2^4 \qquad \text{(Substituted for 16)}$$

We are now in a position to *equate the exponents* (set the exponent from the left side equal to the exponent from the right side).

$$6x - 8 = 4$$

Solve for $x$:

$$6x = 4 + 8$$

$$6x = 12$$

$$x = \frac{12}{6}$$

$$x = 2$$

So, the correct answer is (A).

## • PROBLEM 1-22

Three times the first of three consecutive odd integers is three more than twice the third. What is the second of the three consecutive odd integers?

(A)  7

(D)  13

(B)  9

(E)  15

(C)  11

## SOLUTION:

**(D)**    Three consecutive odd integers can be written as

$x$ = first odd integer

$x + 2$ = second odd integer

$x + 4$ = third odd integer

Now, we must translate the word problem into its mathematical equivalent. Follow closely how each word in the problem is replaced by a number or symbol.

(three) (times) (the first) (is) (three) (more than) (twice) (the third)

3      ×      $x$      =      3      +      2 ×      $(x + 4)$

or,    $3x = 3 + 2(x + 4)$

Using the distributive property we get

$3x = 3 + 2x + 8.$

Collecting like terms on the left gives

$3x - 2x = 3 + 8$

$x = 11$ (first odd integer)

We have just found the first odd integer, $x$, but the problem asks for the *second* odd integer.

second odd integer = $x + 2 = 11 + 2 = 13$

So, the correct answer is (D).

### • PROBLEM 1–23

The length of a rectangle is four more than twice the width. The perimeter of the rectangle is 44 meters. Find the length.

(A)  6 m                  (D)  16 m

(B)  8 m                  (E)  22 m

(C)  11 m

## SOLUTION:

**(D)**   Let $l$ = length

  $w$ = width

Translate the given information into an equation.

  (length) (is) (four) (more than) (twice) (the width)

  $l$   =   4   +   2 ×   $w$

or,   $l = 4 + 2w$                                   (1)

We will soon plug this relation into the perimeter equation.

  perimeter = $44 = l + l + w + w$.

Substituting $(4 + 2w)$ for $l$, from equation (1), gives.

  perimeter = $44 = (4 + 2w) + (4 + 2w) + w + w$

  $= 4 + 2w + 4 + 2w + w + w$

  $= 6w + 8$

or,   $44 = 6w + 8$

Solve for $w$.

  $44 - 8 = 6w$

  $36 = 6w$

  $w = \dfrac{36}{6}$

  $w = 6$

We have just found the width, $w$, but the problem asks for length. To find the length, use equation (1).

  $l = 4 + 2w$

  $= 4 + (2)\,(6)$     (Substituted 6 for $w$)

  $= 4 + 12$

  $= 16$

The length is 16 meters.

So, the correct answer is (D).

## • PROBLEM 1-24

John bought a $250 radio. The salesman gave him a 10% discount. How much did he pay for the radio?

(A) $25                    (D) $225

(B) $125                   (E) $275

(C) $175

## SOLUTION:

**(D)**    John bought the radio at a discount, so he payed *less* than the original price.

First, find the discount.

discount = 10% of original price

$$= .10 \times 250 \qquad \text{(Wrote 10\% in decimal form)}$$

$$= \$25$$

Next, put the value we just found for the discount into the following equation:

(Original price) – (discount) = (price John payed)

(250) – (25) = (price John payed)

$225 = price John payed

So, the correct answer is (D).

## • PROBLEM 1-25

Find the area of the triangle in the figure shown below.

(A) $\dfrac{9\sqrt{3}}{2}$

(B) $9\sqrt{3}$

(C) 18

(D) $18\sqrt{3}$

(E) 24

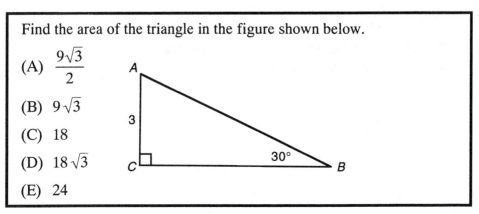

## SOLUTION:

**(A)** *For a triangle:*

$$\text{Area} = \frac{1}{2} \times \text{base} \times \text{height}$$

$$= \frac{1}{2} bh$$

To use this formula, we need to know $h$ and $b$. We do know $h$ ($h = 3$, from the diagram).

However, we cannot apply the formula until we find $b$.

*To find b (length of the base):*

(a) Note that the triangle pictured is a $30 - 60 - 90$ triangle.

(b) Recall that the lengths of the sides of a $30 - 60 - 90$ triangle are in the ratio $1 : \sqrt{3} : 2$. (This gives just the ratio of their lengths, not their actual lengths.)

(c) To find the actual lengths of the sides for the pictured triangle, multiply the ratio through by 3 (since our smallest side has a length of 3, not a length of 1).

$$3 \times (1 : \sqrt{3} : 2) = (3 \times 1) : (3 \times \sqrt{3}) : (3 \times 2)$$

$$= 3 : 3\sqrt{3} : 6$$

This tells us that for our particular triangle:

length of the smaller side = 3

length of the larger side (the base) = $3\sqrt{3}$

length of the hypotenuse = 6

Thus, the quantity we were trying to find, b, is equal to $3\sqrt{3}$.

(We do not need to know the hypotenuse for this problem, but it was shown here for the sake of completeness).

*Find the area:*

$$\text{Area} = \frac{1}{2} bh$$

$$= \frac{1}{2}(3\sqrt{3})(3) \qquad \text{(Substituted for } b \text{ and } h\text{)}$$

$$= \frac{9\sqrt{3}}{2}$$

So, the correct answer is (A).

### • PROBLEM 1-26

I filled $2/3$ of my swimming pool with 1,800 ft$^3$. What is the total capacity of my swimming pool?

(A)  2,400 ft$^3$          (D)  3,600 ft$^3$

(B)  2,700 ft$^3$          (E)  3,200 ft$^3$

(C)  3,000 ft$^3$

## SOLUTION:

**(B)**     We want to find how much water the swimming pool can hold when it is full. This is called the capacity of the swimming pool.

Given:

$$\frac{2}{3} \times (\text{capacity of swimming pool}) = 1,800$$

Solve:

$$(\text{capacity of swimming pool}) = 1,800 \times \frac{3}{2}$$

$$= \frac{1,800 \times 3}{2}$$

$$= 2,700 \text{ ft}^3$$

So, the correct answer is (B).

### • PROBLEM 1-27

A jug will hold $2/3$ gallon of punch. How much punch is in the jug when it is $3/4$ full?

(A)  $\dfrac{5}{7}$ gallon

(B)  $\dfrac{1}{2}$ gallon

(C)  $\dfrac{1}{4}$ gallon

(D)  $\dfrac{2}{3}$ gallon

(E)  $\dfrac{1}{12}$ gallon

### SOLUTION:

**(B)**    Note that a *full* jug contains $2/3$ gallon.

Thus, $3/4$ of a full jug

$$= \frac{3}{4} \times \frac{2}{3} \text{ gallon}$$

(Replaced "full jug" by $2/3$ gallon)

$$= \frac{6}{12} \text{ gallon}$$

$$= \frac{1}{2} \text{ gallon}$$

So, the correct answer is (B).

### • PROBLEM 1-28

If $f(x) = x + 1$, $g(x) = 2x - 3$, and an operation $*$ is defined for all real numbers $a$ and $b$ by the equation $a * b = 2a + b - ab$, then $f(3) * g(4) =$

(A)  −9.

(B)  −7.

(C)  −1.

(D)  0.

(E)  5.

## SOLUTION:

**(B)**   *Here's our plan:*

*Step I*—Find $f(3)$ and $g(4)$.

*Step II*—Use these values to find $f(3) * g(4)$, where $*$ is defined in the problem.

*Step I*

(a)   $f(x) = x + 1$

$f(3) = 3 + 1$        (Substituted 3 for $x$)

So,   $f(3) = 4$

(b)   $g(x) = 2x - 3$

$g(4) = (2)(4) - 3$    (Substituted 4 for $x$)

$g(4) = 8 - 3$

So,   $g(4) = 5$

*Step II*

We want to find $f(3) * g(4)$.

We have a formula for $a * b$.

For use in the formula, assign the following:

$a = f(3) = 4$

$b = g(4) = 5$

Then replace each "$a$" in the formula with 4 and each "$b$" with 5.

$$f(3) * g(4) = a * b$$
$$= 2a + b - ab$$
$$= 2(4) + 5 - 4 \times 5 \quad \text{(Substituted } a = 4, b = 5)$$
$$= 8 + 5 - 20$$
$$= 13 - 20$$
$$= -7$$

So, the correct answer is (B).

• **PROBLEM 1–29**

What is the value(s) of $x$ in the equation $(4x - 3)^2 = 4$?

(A) $\dfrac{5}{4}$

(B) $\dfrac{1}{4}$

(C) $\dfrac{5}{4}, \dfrac{1}{4}$

(D) $\dfrac{1}{4}, \dfrac{5}{2}$

(E) $\dfrac{5}{2}, \dfrac{1}{5}$

## SOLUTION:

**(C)** We are given the equation $(4x - 3)^2 = 4$.

Take the square root of both sides to get the following two equations:

$$\sqrt{(4x + 3)^2} = \sqrt{4} \qquad\qquad \sqrt{(4x + 3)^2} = -\sqrt{4}$$

$$4x - 3 = 2 \qquad\qquad\qquad 4x - 3 = -2$$

$$4x = 2 + 3 \qquad\qquad\qquad 4x = -2 + 3$$

$$x = \frac{5}{4} \qquad \text{or} \qquad\qquad x = \frac{1}{4}$$

Thus, there are two different values of $x$ which, if plugged into the original equation $(4x - 3)^2 = 4$, will make the equation true.

The correct answer is (C).

• **PROBLEM 1–30**

Fifteen percent of what number is 60?

(A) 9

(B) 51

(C) 69

(D) 200

(E) 400

## SOLUTION:

**(E)**    Translate the given information into an equation.

(fifteen percent) (of) (what number) (is) 60

$.15 \times x = 60$

or,    $.15\, x = 60$

Solve for $x$.

$$x = \frac{60}{.15}$$

$$= 400$$

So, the correct answer is (E).

### • PROBLEM 1–31

$10^3 + 10^5 =$

(A)  $10^8$

(B)  $10^{15}$

(C)  $20^8$

(D)  $2^{15}$

(E)  $101{,}000$

## SOLUTION:

**(E)**

$$10^3 + 10^5 = (10)(10)(10) + (10)(10)(10)(10)(10)$$

$$= 1{,}000 + 100{,}000$$

$$= 101{,}000$$

So, the correct answer is (E).

### • PROBLEM 1-32

If $a = 4$ and $b = 7$, then $\dfrac{a + \frac{a}{b}}{a - \frac{a}{b}}$

(A) $\dfrac{3}{4}$.

(D) $\dfrac{4}{3}$.

(B) $\dfrac{3}{7}$.

(E) $\dfrac{7}{3}$.

(C) 1.

## SOLUTION:

**(D)**    Every place you see an "*a*," substitute 4.

Every place you see a "*b*," substitute 7.

$$\frac{a + \frac{a}{b}}{a - \frac{a}{b}}$$

(Given)

$$= \frac{4 + \frac{4}{7}}{4 - \frac{4}{7}}$$

(Substituted $a = 4$ and $b = 7$)

$$= \frac{\left[4 \times \left(\frac{7}{7}\right)\right] + \frac{4}{7}}{\left[4 \times \left(\frac{7}{7}\right)\right] - \frac{4}{7}}$$

(In the brackets, we are finding how many sevenths the number 4 is equal to.)

$$= \frac{\frac{28}{7} + \frac{4}{7}}{\frac{28}{7} - \frac{4}{7}}$$

(Inside the brackets, we found that $4 = {}^{28}/_{7}$.)

$$= \frac{\frac{32}{7}}{\frac{24}{7}}$$

$$= \frac{32}{7} \times \frac{7}{24}$$

(Used the fact that $\dfrac{\frac{a}{b}}{\frac{c}{d}} = \dfrac{a}{b} \times \dfrac{d}{c}$)

$$= \frac{32}{24}$$

$$= \frac{4}{3}$$

So, the correct answer is (D).

## • PROBLEM 1–33

The greatest area that a rectangle whose perimeter is 52 m can have is

(A)  12 m².                     (D)  168 m².

(B)  169 m².                    (E)  52 m².

(C)  172 m².

## SOLUTION:

**(B)**  *You need these facts to understand which rectangle to use:*

(1)  Some rectangles are long and thin; others are closer to being square; (also a square is itself a rectangle).

(2)  It is possible to find several differently shaped rectangles which have the same perimeter, but different area.

(3)  It turns out that a *square* encompasses a greater area than any longer, thinner rectangle with the same perimeter (square gives maximum area).

(4)  The problem asks for "greatest area," so our rectangle must be a square.

*Find the length of a side:*

$$\text{length of one side} = \frac{\text{perimeter}}{4}$$

(For a square, lengths of sides are equal; divide perimeter into four equal parts.)

$$= \frac{52}{4}$$

$$= 13 \text{ m}$$

*Find the area:*

$$\text{Area} = \text{length} \times \text{width}$$

$$= 13 \times 13$$

(Length and width are equal for a square.)

$$= 169 \text{ m}^2$$

So, the correct answer is (B).

## • PROBLEM 1–34

The number 120 is separated into two parts. The larger part exceeds three times the smaller by 12. The smaller part is

(A)  27.

(B)  33.

(C)  15.

(D)  39.

(E)  29.

## SOLUTION:

**(A)**     Separate the number 120 into two parts.

$$x = \text{larger part}$$

$$120 - x = \text{smaller part}$$

(larger part + smaller part = 120; so, 120 – larger part = smaller part)

Translate the given information into an equation.

(The larger part) = (three times) (the smaller part) + 12

$$x \qquad = \qquad 3 \times \qquad (120 - x) \qquad + 12$$

or,     $x = 3(120 - x) + 12$

Using the distributive property gives

$$x = (3)(120) - 3x + 12$$

$$x = 360 - 3x + 12$$

Move all terms containing $x$ to the left side and all pure numbers to the right side to give

$$x + 3x = 360 + 12$$

$$4x = 372$$

$$x = \frac{372}{4}$$

$$x = 93 \qquad \text{(Larger part)}$$

The question asks for the *smaller* part.

$$\text{smaller part} = 120 - x$$

$$= 120 - 93$$

(Substituted for $x$, from above)

$$= 27$$

So, the correct answer is (A).

## • PROBLEM 1-35

Of the following relations, the ones that are functions are:

I. $\dfrac{x^2}{81} - \dfrac{y^2}{16} = 3$

II. $x^2 + \left| \dfrac{\sqrt{y^2}}{3} \right| = 3y$

III. $y = \sqrt{3}x$

(A) I only

(B) I and III only.

(C) II only.

(D) I, II, and III.

(E) II and III only.

## SOLUTION:

**(E)** *To determine if a relation is a function:*

(1) Solve for $y$ if possible.

(2) Use the following test:

If there is *only one y* value for each *x* value: *Yes,* a function.

If there is *more than one y* value for any *x* value: *No,* not a function.

*Now, let's test each of the given relations:*

I. $\dfrac{x^2}{81} - \dfrac{y^2}{16} = 3$

It would be too time consuming to solve this exactly, and anyway it is not necessary. We are only interested in the relationship between *x* and *y*, so we don't care what the constants are.

We rewrite the equation, showing the essential features.

$x^2 - y^2 =$ some known constants

We'll get tired of writing "some known constants," so let's just write "*k.*"

$$x^2 - y^2 = k$$
$$-y^2 = k - x^2$$
$$y^2 = x^2 - k$$
$$y = \pm\sqrt{x^2 - k}$$

Imagine now that you plug in an *x* value on the right side. You'll see that there is *more than one* value of *y* for a given *x.*

These *y* values are

$$y = +\sqrt{x^2 - k} \text{ and } y = -\sqrt{x^2 - k}.$$

Thus, part I is *not* a function.

II. $x^2 + \left| \dfrac{\sqrt{y^2}}{3} \right| = 3y$

This one could trick you. If you spotted the $x^2$ and $y^2$, you might mistakenly assume that this is just like the last problem. It's not. Here the $\sqrt{y^2}$ simplifies down to just *y* to give

$$x^2 + \left| \dfrac{y}{3} \right| = 3y.$$

Note that $y$ is positive (or zero), since we obtained it by taking a square root. (The symbol $\sqrt{\phantom{x}}$ indicates the *positive* square root.) Recall also that the absolute value of a *positive* number is that number.

Thus, $\left|\dfrac{y}{3}\right| = \dfrac{y}{3}$ for this problem.

Removing the absolute value sign gives:

$$x^2 + \frac{y}{3} = 3y$$

$$3y - \frac{y}{3} = x^2$$

$$\frac{9y}{3} - \frac{y}{3} = x^2$$

$$\frac{8y}{3} = x^2$$

$$y = \frac{3}{8}x^2$$

Let's apply our function test. Imagine that you plug in some value of $x$ on the right side. This value will be squared and then multiplied by $3/8$ to give a *single* $y$ value. Since there is *only one* $y$ value for a given $x$, part II is a function.

III.    $y = \sqrt{3}x$

This one is easy—it's already in the form we need. Each $x$ value gets multiplied by $\sqrt{3}$ to give *one* $y$ value. Part III is thus a function.

We have shown that only parts II and III are functions.

So, the correct answer is (E).

Note: If you were to draw the graphs for these equations you would find

Part I—a hyperbola

Part II—a parabola

Part III—a straight line

## • PROBLEM 1–36

Two pounds of pears and one pound of peaches cost $1.40. Three pounds of pears and two pounds of peaches cost $2.40. How much is the combined cost of one pound of pears and one pound of peaches?

(A) $2.00

(B) $1.50

(C) $1.60

(D) $.80

(E) $1.00

## SOLUTION:

**(E)** To solve this problem we must set up a simultaneous equation.

Let $x$ = the cost of one pound of pears

$2x$ = the cost of two pounds of pears

Let $y$ = the cost of one pound of peaches

$2x + y = 1.40$          (1)

Similarly, we have

$3x$ = the cost of three pounds of pears

$2y$ = the cost of two pounds of peaches

$3x + 2y = 2.40$         (2)

Equations (1) and (2) are solved simultaneously as follows:

$2x + 2x + y = 1.40$       (1)

$3x + 3x + 2y = 2.40$      (2)

In equation (1) we solve for $y$ in terms of $x$.

$2x + y = 1.40$

$y = 1.40 - 2x$

We then substitute $y = 1.40 - 2x$ for $y$ in equation (2).

$3x + 2y = 2.40$

$3x + 2(1.40 - 2x) = 2.40$

Solving this equation for $x$ we obtain the following.

$3x + 2(1.40 - 2x) = 2.40$

$$3x + 2.80 - 4x = 2.40$$

$$-1x = -.40$$

$$x = .40$$

Thus, one pound of pears cost $.40.

Substituting $.40 for the value of $x$ in equation (1), we can obtain the value of $y$.

$$2 (.40) + y = 1.40$$

$$.80 + y = 1.40$$

$$y = .60$$

Therefore, the cost of one pound of pears and one pound of peaches is .40 + .60 = 1.00.

The correct choice is (E).

## • PROBLEM 1–37

The length of a rectangle is $6L$ and the width is $4W$. What is the perimeter?

(A)  $12L + 8W$          (D)  $20LW$

(B)  $12L^2 + 8W^2$      (E)  $24LW$

(C)  $6L + 4W$

## SOLUTION:

**(A)**    The perimeter of a rectangle is equal to the sum of its sides.

The perimeter is equal to 2 × (length) plus 2 × (width).

Therefore, the perimeter equals

$$2 (6L) + 2 (4W) = 12L + 8W.$$

The correct answer is (A).

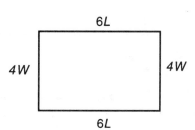

## • PROBLEM 1–38

Using order of operations, solve: $3 \times 6 - 12 \div 2$.

(A) $-9$.

(B) 3.

(C) 6.

(D) 12.

(E) 18.

**SOLUTION:**

**(D)** According to the order of operations, we perform all multiplication and division first from left to right. Then additions and subtractions are performed from left to right.

$$3 \times 6 - 12 \div 2 = 18 - 12 \div 2$$

$$= 18 - 6$$

$$= 12$$

The correct answer is (D).

## • PROBLEM 1–39

Change $4^5/_6$ to an improper fraction.

(A) $\dfrac{5}{24}$

(B) $\dfrac{9}{6}$

(C) $\dfrac{29}{6}$

(D) $\dfrac{30}{4}$

(E) $\dfrac{120}{6}$

**SOLUTION:**

**(C)** To change $4^5/_6$ to an improper fraction, we multiply the whole number by the denominator and add the numerator.

$$\frac{4(6)+5}{6} = \frac{29}{6}$$

The correct answer is (C).

## • PROBLEM 1-40

Find the sum of

$$5\frac{3}{4}, 2\frac{11}{16}, \text{ and } 7\frac{1}{8}.$$

(A) $14\frac{8}{17}$

(D) $15\frac{15}{28}$

(B) $14\frac{15}{16}$

(E) $15\frac{9}{16}$

(C) $15\frac{1}{2}$

## SOLUTION:

**(E)** In order to find the sum of these numbers, we must first find a common denominator for the fractional component.

$$\frac{3}{4} = \frac{12}{16}$$

$$\frac{11}{16} = \frac{11}{16}$$

$$\frac{1}{8} = \frac{2}{16}$$

16 is the lowest common denominator (LCD). Therefore,

$$5\frac{3}{4} = 5\frac{12}{16}$$

$$2\frac{11}{16} = 2\frac{11}{16}$$

$$7\frac{1}{8} = 7\frac{2}{16}$$

We then add the whole numbers and the fractional components.

$$5\frac{12}{16} + 2\frac{11}{16} + 7\frac{2}{16} = 14\frac{25}{16}$$

$14\frac{25}{16}$ can be reduced to $15\frac{9}{16}$.

The correct answer is (E).

## • PROBLEM 1–41

A counting number with exactly two different factors is called a prime number. Which of the following pairs of numbers are consecutive prime numbers?

(A)  27 and 29

(B)  31 and 33

(C)  35 and 37

(D)  37 and 29

(E)  41 and 43

### SOLUTION:

**(E)**   A prime number is a number that is divisible by one and itself only.

Analyzing the choices, we can see that in choices (A), (B), and (C) at least one of the numbers in each pair is not a prime number.

In choice (A), $27 = 9 \times 3$, 29 is prime.

In choice (B), $33 = 3 \times 11$, 31 is prime.

In choice (C), $35 = 5 \times 7$, 37 is prime.

In choice (D), both numbers are prime, but they are not consecutive.

In answer (E), 41 and 43 are both prime numbers and consecutive numbers that are prime.

## • PROBLEM 1–42

What part of three-fourths is one-tenth?

(A)  $\dfrac{1}{8}$

(B)  $\dfrac{15}{2}$

(C)  $\dfrac{2}{15}$

(D)  $\dfrac{3}{40}$

(E)  None of the above.

## SOLUTION:

**(C)**    We must translate the statement into an equation:

(What part) (of) (three-fourths) (is) (one-tenth)

$$x \quad \times \quad \frac{3}{4} \quad = \quad \frac{1}{10}$$

or,    $\dfrac{3}{4}x = \dfrac{1}{10}$

Solve for $x$,

$$x = \frac{\frac{1}{10}}{\frac{3}{4}} = \frac{1}{10} \times \frac{4}{3}$$

Dividing by a fraction is the same as multiplying by its reciprocal. We multiply $1/10$ by the reciprocal of $3/4$ which is $4/3$.

The correct choice is (C).

## • PROBLEM 1–43

Change the fraction $7/8$ to a decimal.

(A) .666          (D) .875

(B) .75           (E) 1.142

(C) .777

## SOLUTION:

**(D)**    To change a fraction to a decimal divide the numerator, 7, by the denominator, 8. Add a decimal point after the 7 and the necessary zeros and continue dividing.

$$\begin{array}{r} .875 \\ 8\overline{\smash)7.000} \\ \underline{-64} \\ .60 \\ \underline{-56} \\ .40 \\ \underline{-40} \\ 0 \end{array}$$

The correct choice is (D).

## • PROBLEM 1–44

Twelve more than twice a number is 31 less than three times the number. Find the number.

(A) – 43                    (D) 19

(B) – 19                    (E) 43

(C) – 9

## SOLUTION:

**(E)**     Let $x$ = the number

We must convert the sentence into an equation.

$2x + 12 = 3x - 31$

Now we solve for $x$ by subtracting $2x$ from both sides.

$-2x + 2x + 12 = 3x - 31 - 2x$

$12 = x - 31$

Adding 31 to both sides we get

$43 = x.$

The correct answer is (E).

## • PROBLEM 1–45

The area of Jane's living room is 48 m². The length of the room is 2 m more than the width. What is the length?

(A)  4 m                    (D)  10 m

(B)  6 m                    (E)  Cannot be determined.

(C)  8 m

## SOLUTION:

**(C)**     Let $w$ = width of the room

$l$ = length of the room

Since the length is 2 more than the width,

$$l = (w + 2)$$

Using the formula for area (area = length × width) and substituting the above information, we get

$$(w + 2) w = 48$$

$$w^2 + 2w = 48$$

$$w^2 + 2w - 48 = 0$$

Factoring this equation, we obtain

$$(w + 8) (w - 6) = 0.$$

Setting each partial product to zero,

$$w + 8 = 0 \qquad w - 6 = 0$$

$$w = -8 \qquad w = 6$$

Although there are two possible solutions for $w$, there cannot be a negative value for a room dimension. Therefore, $w = 6$ and $l = w + 2 = 6 + 2 = 8$.

The correct choice is (C).

## • PROBLEM 1–46

Ron saves $38 in four weeks. How many weeks will it take Ron to save $152 at the same rate?

(A)  12　　　　　　　　(D)  18

(B)  14　　　　　　　　(E)  20

(C)  16

## SOLUTION:

**(C)**　　Ron saves $38 in 4 weeks.

We can find the number of weeks it will take Ron to save $152 by setting up a proportion.

$$\frac{\$38}{4 \text{ weeks}} = \frac{\$152}{x \text{ weeks}}$$

We can solve for $x$ by cross multiplying.

$$38x = 4(152)$$

$$38x = 608$$

$$x = \frac{608}{38}$$

$$x = 16$$

The correct answer is (C).

## • PROBLEM 1–47

Find the area of a right triangle with a hypotenuse of 17 cm and a base of 8 cm.

(A)  45 cm$^2$         (D)  120 cm$^2$

(B)  60 cm$^2$         (E)  138 cm$^2$

(C)  68 cm$^2$

## SOLUTION:

**(B)**     The area of a triangle is $\frac{1}{2}$ base × height. The base is 8 and the hypotenuse is 17. To find the height, we use the Pythagorean Theorem.

$$h^2 + 8^2 = 17^2$$

$$h^2 + 64 = 289$$

$$h^2 = 225$$

$$h = \sqrt{225}$$

$$h = 15$$

Now that we have $h = 15$, we can substitute this value and find the area of the triangle.

$$\text{Area} = \frac{1}{2} \text{ (base) (height)}$$

$$\text{Area} = \frac{1}{2} \text{ (8 cm) (15 cm)}$$

Area = 60 cm$^2$

The correct choice is (B).

## • PROBLEM 1-48

Find the cost of seeding a lawn shaped like a parallelogram with a base of 6 m and a height of 5 m. One kilogram of grass seed covers 15 m$^2$ and costs $7.50.

(A) $3.75

(D) $15.00

(B) $7.50

(E) $22.50

(C) $11.25

## SOLUTION:

**(D)**

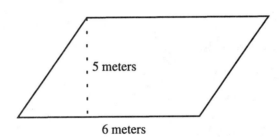

5 meters

6 meters

The area of a parallelogram = (base) (height)

Area = (6 m) (5 m) = 30 m$^2$

One kilogram of grass seed covers 15 m$^2$ and costs $7.50. We would need two kilograms of grass seed to cover 30 m$^2$. This would cost 2 × $7.50 = $15.00.

The correct choice is (D).

## • PROBLEM 1–49

If the measure of an angle exceeds its complement by 40°, then its measure is

(A) 65°.

(B) 50°.

(C) 45°.

(D) 40°.

(E) 30°.

**SOLUTION:**

**(A)** Two angles that are complementary add up to 90°.

Let $x$ = the angle

$x - 40$ = the complement

$$x + (x - 40) = 90$$

$$2x - 40 = 90$$

$$2x = 90 + 40$$

$$2x = 130$$

$$x = 65°$$

The correct choice is (A).

## • PROBLEM 1–50

Two dice are thrown, one red and one green. The probability that the number on the red exceeds the number showing on the green by exactly two is

(A) $\frac{1}{18}$.

(B) $\frac{1}{4}$.

(C) $\frac{1}{9}$.

(D) $\frac{1}{36}$.

(E) $\frac{1}{24}$.

## SOLUTION:

**(C)**    If two dice are thrown, there are six possibilities for each die (1, 2, 3, 4, 5, or 6). Therefore, there are $6^2 = 36$ possibilities for the total number of combinations on a pair of dice.

If the number on the red die exceeds the number on the green die by two, we have the following combinations possible.

| Red | Green | Combinations |
|-----|-------|--------------|
| 2 | 2 − 2 = 0  0 | |
| 3 | 3 − 2 = 1  1 | 3, 1 |
| 4 | 4 − 2 = 2  2 | 4, 2 |
| 5 | 5 − 2 = 3  3 | 5, 3 |
| 6 | 6 − 2 = 4  4 | 6, 4 |

There are only four combinations out of 36 that satisfy the event that the number on the red die exceeds the number on the green die by 2. Hence, the probability for this event to happen is $^4/_{36} = ^1/_9$.

The correct choice is (C).

### • PROBLEM 1-51

$$\sqrt{X\sqrt{X\sqrt{X}}} =$$

(A) $X^{7/8}$            (D) $X^{3/4}$

(B) $X^{7/4}$            (E) $X^{15/8}$

(C) $X^{15/16}$

## SOLUTION:

**(A)**    The best way to handle this problem is to simplify in steps.

$$\sqrt{X\sqrt{X\sqrt{X}}} =$$

Since $\sqrt{X} = X^{1/2}$ then

$$\sqrt{X\sqrt{X \times \sqrt{X}}} = \sqrt{X\sqrt{X \times X^{1/2}}} .$$

We can simplify $\sqrt{X \times X^{1/2}}$ by using 2 as a common denominator for the exponents.

Therefore,

$$\sqrt{X \times X^{1/2}} = \sqrt{X^{2/2} \times X^{1/2}}$$
$$= \sqrt{X^{3/2}}$$
$$= \left(X^{3/2}\right)^{1/2}$$
$$= X^{3/4}$$

Going back to the original problem, we have

$$\sqrt{X\sqrt{X^{3/2}}} = \sqrt{X \times X^{3/4}}.$$

Using 4 as a common denominator of the exponents, we can simplify again.

$$\sqrt{X^{4/4} \times X^{3/4}} = \sqrt{X^{7/4}}$$
$$= (X^{7/4})^{1/2}$$
$$= X^{7/8}$$

The correct choice is (A).

## • PROBLEM 1–52

A line segment is drawn from the point (3, 5) to the point (9, 13). What are the coordinates of the midpoint of the line segment?

(A)  (9, 6)          (D)  (12, 18)

(B)  (6, 9)          (E)  (6, 8)

(C)  (3, 4)

### SOLUTION:

**(B)**    If two points on a line segment are $(x_1, y_1)$ and $(x_2, y_2)$, then the midpoint between these two points is given by

$$\left(\frac{x_1 + x_2}{2}, \frac{y_1 + y_2}{2}\right).$$

If we consider the line segment connecting the points (3, 5) to (9, 13), then

$$x_1 = 3 \qquad x_2 = 9$$

$$y_1 = 5 \qquad y_2 = 13$$

Therefore, the midpoint of the line segment is

$$\left(\frac{3+9}{2}, \frac{5+13}{2}\right) = \left(\frac{12}{2}, \frac{18}{2}\right)$$
$$= (6, 9)$$

The correct choice is (B).

### • PROBLEM 1-53

The number missing in the series 2, 6, 12, 20, $x$, 42, 56 is

(A)  36.  (D)  38.

(B)  24.  (E)  40.

(C)  30.

## SOLUTION:

**(C)**   We note the following pattern arithmetically by considering the following differences:

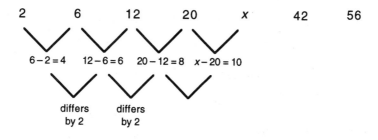

Then $x - 20 = 10$ in keeping with the pattern

Solving for $x$,

$$x - 20 = 10$$

$$x = 30$$

To check the answer, analyze the pattern with $x = 30$.

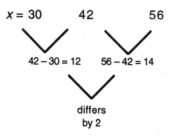

We see that the pattern continues. Therefore, the missing piece is 30.

The correct choice is (C).

### • PROBLEM 1-54

A truck contains 150 small packages, some weighing 1 kg each and some weighing 2 kg each. How many packages weighing 2 kg each are in the truck if the total weight of all the packages is 264 kg?

(A) 36

(B) 52

(C) 88

(D) 124

(E) 114

## SOLUTION:

**(E)**   Let $(x)$ = the number of packages weighing 2 kilograms each.

$(150 - x)$ = the number of packages weighing 1 kilogram each.

$2(x)$ = total weight of all 2 kilogram packages.

$1(150 - x)$ = total weight of 1 kilogram packages.

The total weight of all packages is 264.

$$2x + (150 - x) = 264$$

Combining like terms, we can simplify.

$$2x + 150 - x = 264$$

$$x + 150 = 264$$

$$x = 264 - 150$$

$$x = 114$$

There are 114 packages weighing 2 kilograms each.

The correct choice is (E).

## • PROBLEM 1–55

The solution of the equation $4 - 5(2y + 4) = 4$ is

(A) $-\dfrac{2}{5}$.

(D) $-2$.

(B) 8.

(E) None of these.

(C) 4.

## SOLUTION:

**(D)**

$$4 - 5(2y + 4) = 4$$

To solve the problem we first apply the distribution law to obtain the following:

$$4 - 10y - 20 = 4$$

We can further simplify by combining numerical terms on the left side.

$$-10y - 16 = 4$$

Combining numerical terms and solving for $y$, we get

$$-10y = 20$$

$$\frac{-10y}{-10} = \frac{20}{-10}$$

$$y = -2$$

The correct answer is (D).

### • PROBLEM 1–56

What is the product of $(\sqrt{3}+6)$ and $(\sqrt{3}-2)$?

(A) $9+4\sqrt{3}$         (D) $-9+2\sqrt{3}$

(B) $-9$         (E) $9$

(C) $-9+4\sqrt{3}$

## SOLUTION:

**(C)**

$$(\sqrt{3}+6)(\sqrt{3}-2) = (\sqrt{3}+6)(\sqrt{3})-(\sqrt{3}+6)(2)$$
$$= \sqrt{9}+6\sqrt{3}-2\sqrt{3}-12$$
$$= 3+6\sqrt{3}-2\sqrt{3}-12$$
$$= -9+4\sqrt{3}$$

The correct choice is (C).

### • PROBLEM 1–57

Two cyclists start toward each other from two towns that are 135 miles apart. One cyclist rides at 15 mph and the other rides at 12 mph. In how many hours will they meet?

(A) 3         (D) 11

(B) 5         (E) 15

(C) 9

## SOLUTION:

**(B)** Let $x$ = the number of hours it takes for the two cyclists to meet

Distance = rate × time

The cyclist who rides at 15 mph will travel a total of $15x$ miles.

The cyclist who rides at 12 mph will travel a total of $12x$ miles.

The total distance traveled is 135 miles.

We therefore add the distance traveled by both cyclists to obtain

$$12x + 15x = 135$$

$$27x = 135$$

$$x = \frac{135}{27}$$

$$x = 5$$

Therefore, the two cyclists will meet in 5 hours.

The correct choice is (B).

## • PROBLEM 1–58

In rhombus *ABCD,* which of the following are true?

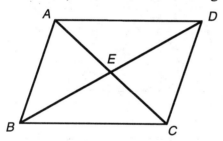

I.   $\angle BAE$ and $\angle ECD$ are congruent.

II.  $\angle ADE$ and $\angle CDE$ are congruent.

III. $\angle ABE$ and $\angle ADE$ are congruent.

(A)  I only.

(B)  II only.

(C)  I and II only.

(D)  I and III only.

(E)  I, II, and III.

## SOLUTION:

**(E)**    A rhombus is a parallelogram in which all four sides are the same length and the opposite sides are parallel. The diagonals of a rhombus bisect its angles. In the diagram sides $\overline{BA}$ and $\overline{CD}$ are parallel.

I. $\angle BAE$ and $\angle ECD$ are congruent since they are alternate interior angles of the parallel sides of the rhombus with $\overline{AC}$ as the transversal.

II. ∠ADE and ∠CDE are congruent since the diagonals of a rhombus bisect the angles of a rhombus.

III. ∠ABE and ∠ADE are congruent since $\overline{AB}$ is congruent to $\overline{AD}$ where $\overline{AB}$ and $\overline{AD}$ are sides of the rhombus. Triangle ABD is isosceles. Therefore the angles opposite these congruent sides are congruent.

The correct choice is (E).

## • PROBLEM 1-59

If the length of segment $\overline{EB}$, base of triangle EBC, is equal to $1/4$ the length of segment $\overline{AB}$ ($\overline{AB}$ is the length of rectangle ABCD), and the area of triangle EBC is 12 square units, find the area of the shaded region.

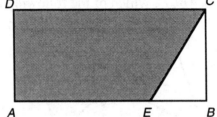

(A)  24 square units

(B)  96 square units

(C)  84 square units

(D)  72 square units

(E)  120 square units

## SOLUTION:

**(C)**  Let $(\overline{AB})$ represent the length of segment $\overline{AB}$. Then the length of rectangle ABCD is equal to $(\overline{AB})$, and its width is $(\overline{BC})$.

The area of the shaded region is equal to the area of rectangle ABCD minus the area of triangle EBC.

Recall that the area of a rectangle is equal to the product of its length and its width. Thus,

Area of rectangle $ABCD = (\overline{AB})(\overline{BC})$.

The area of a triangle is equal to $1/2\, bh$. Thus,

Area of triangle $EBC = 1/2\,(\overline{EB})(\overline{BC})$

But $(\overline{EB}) = 1/4\,(\overline{BC})$. Hence

Area of triangle $EBC = \dfrac{1}{2}\left(\dfrac{1}{4}(\overline{AB})\right)(\overline{BC})$

$$= \frac{1}{8} (\overline{AB}) (\overline{BC})$$

Since the area of triangle $EBC$ is equal to 12 square units, we have

$$\frac{1}{8} (\overline{AB}) (\overline{BC}) = 12$$

or $(\overline{AB}) (\overline{BC}) = 96$.

But, $(\overline{AB}) (\overline{BC})$ is the area of rectangle $ABCD$. Hence, area of rectangle $ABCD = 96$ square units.

Thus, area of shaded region equals

96 – 12 = 84 square units.

The correct answer is (C).

## • PROBLEM 1–60

Suppose the average of two numbers is $WX$. If the first number is $X$, what is the other number?

(A) $WX - X$          (D) $WX - 2X$

(B) $2WX - W$         (E) $2WX - X$

(C) $W$

## SOLUTION:

**(E)**    The average of two numbers is obtained by adding the two numbers together and dividing by 2.

We are given that the average of two numbers is $WX$. We are given that $X$ is the first number. We can let $Y$ represent the second number in the average of the two numbers.

$$\frac{X + Y}{2} = WX$$

Solving for $Y$, we get

$$X + Y = 2 (WX)$$

$$Y = 2WX - X$$

The correct choice is (E).

## • PROBLEM 1–61

Find the slope of a line whose equation is $y = -6x + 3$.

(A) $-\dfrac{1}{6}$  (D) $-6$

(B) $6$  (E) $-3$

(C) $3$

**SOLUTION:**

**(D)**  The slope of a line in the form $y = mx + b$ is given by the coefficient of $x$.

If $y = -6x + 3$, we can see that the slope $m$ is equal to $-6$.

The correct choice is (D).

## • PROBLEM 1–62

Solve for the value of $y$:

$3x + 2y = 12$

$2x - 2y = 8$

(A) $0$  (D) $4$

(B) $2$  (E) $5$

(C) $3$

**SOLUTION:**

**(A)**

$3x + 2y = 12$

$2x - 2y = 8$

If we add these two equations together, we get

$3x + 2y = 12$

$\underline{2x - 2y = \phantom{0}8}$

$5x + \phantom{0}0 = 20$

$$5x = 20$$

$$x = 4$$

To find the value of $y$, we substitute 4 for $x$ in the equation above and solve for $y$.

$$3x + 2y = 12$$

$$3(4) + 2y = 12$$

$$12 + 2y = 12$$

$$2y = 0$$

$$y = 0$$

The correct answer is (A).

## • PROBLEM 1–63

One side of a rectangle is twice the length of the other side, and the perimeter is 36. Find the area of the rectangle.

(A) 48　　　　　　　　　　(D) 128

(B) 72　　　　　　　　　　(E) 144

(C) 90

## SOLUTION:

**(B)**　　Let $x$ = the length of the rectangle

$2x$ = the length of the other side

We are given that the perimeter of the rectangle is equal to 36.

The perimeter is equal to the sum of the length of its sides. Therefore,

$$x + 2x + x + 2x = 36$$

$$6x = 36$$

$$x = 6$$

One side of the rectangle is 6. The other side is $2x = 12$.

To obtain the area of a rectangle, we must multiply the length of one side by the length of the other side.

Area = $x (2x) = 2x^2 = 2(6)^2 = 2(36) = 72$

The correct choice is (B).

## • PROBLEM 1–64

How long of a metal bar do you need to make a basketball hoop with a diameter of 48 cm?

(A)  75.36 cm            (D)  602.88 cm

(B)  150.72 cm           (E)  15,072 cm

(C)  301.44 cm

## SOLUTION:

**(B)**    We are interested in finding the circumference of the hoop. We are given that the diameter is 48 cm.

The circumference of a circle is found by using the equation

Circumference = $\pi d$.

The value of $\pi = 3.14$. Therefore,

Circumference = $\pi (48)$

$$= (3.14) (48)$$

$$= 150.72$$

(B) is the correct choice.

## • PROBLEM 1–65

The length of a rectangle is four more than twice the width. The perimeter of the rectangle is 44 meters. Find the length.

(A)  6 m            (D)  16 m

(B)  8 m            (E)  22 m

(C)  11 m

## SOLUTION:

**(D)**    Let $w$ = width of the rectangle

$2w + 4$ = length of the rectangle

We are given that the perimeter of the rectangle is 44. Therefore,

Perimeter = $2(w) + 2(l)$

$2(w) + 2(2w + 4) = 44.$

We can now solve for $w$.

$2w + 4w + 8 = 44$

$6w + 8 = 44$

$6w = 36$

$w = 6$

In order to find the length, we substitute 6 into the expression $2w + 4$.

$2w + 4 = 2(6) + 4 = 16$

The length is 16.

The correct choice is (D).

## • PROBLEM 1-66

Which of the following represent functions?

I.

II.

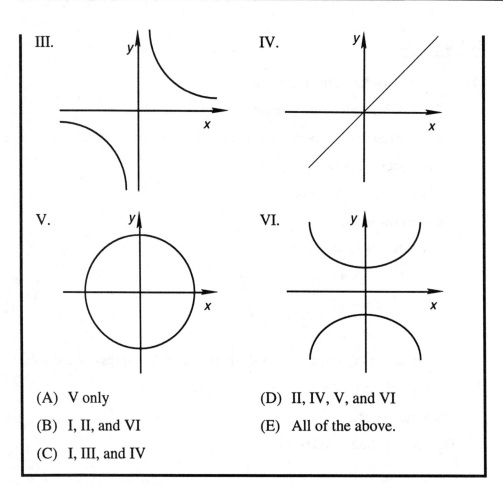

(A)  V only

(B)  I, II, and VI

(C)  I, III, and IV

(D)  II, IV, V, and VI

(E)  All of the above.

## SOLUTION:

**(C)**  *Quick and easy function test to use when the graph is given to you:*

1)  Draw a vertical line through the graph.

2)  If the line you drew intersects the graph in *only one* point: *Yes,* it is a function.

If the line you drew intersects the graph in *more than one* point: *No,* it is not a function.

I.  The line intersects the graph in *one* point.

Yes, it is a function.

[Parabola opening upward]

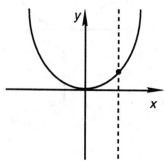

II.  The line intersects the graph in *two* points.

No, it is not a function.

[Parabola opening sideways]

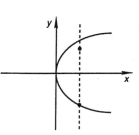

III.  The line intersects the graph in *one* point.

Yes, it is a function

[Hyperbola of the form $xy = c$]

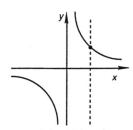

IV.  The line intersects the graph in *one* point.

Yes, it is a function.

[Straight line]

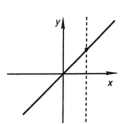

V.  The line intersects the graph in *two* points.

No, it is not a function.

[Circle]

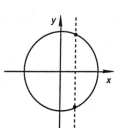

VI.  The line intersects the graph in *two* points.

No, it is not a function.

[Hyperbola of the form $\dfrac{y^2}{a^2} - \dfrac{x^2}{b^2} = 1$]

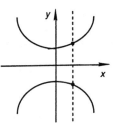

So, the correct answer is (C).

Note: When your vertical line intersects the graph in two points, this means there are *two y-values* for a given *x*-value (not a function).

*To summarize:*

1) *Functions*—parabolas opening up or down; hyperbolas of the form $xy = c$; straight lines (except vertical ones).

2) *Not functions*—parabolas opening right or left; hyperbolas of the form

$$\frac{x^2}{a^2} - \frac{y^2}{b^2} = 1, \text{ or } \frac{y^2}{a^2} - \frac{x^2}{b^2} = 1;$$

circles; ellipses.

## • PROBLEM 1-67

In the figure shown below, find $x$.

(A) 50

(B) 80

(C) 120

(D) 130

(E) 150

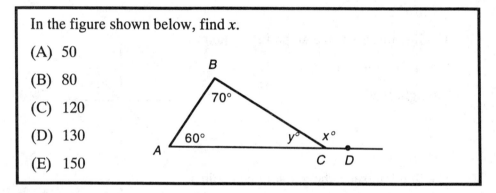

## SOLUTION:

**(D)**     The sum of the angles of a triangle is 180°.

$$\angle A + \angle B + y = 180°$$

Since $\angle A = 60$ and $\angle B = 70$, we can substitute these values in the above equation.

$$60 + 70 + y = 180°$$

Solving for $y$, we find that $y = 50°$.

$\angle BCD$ is adjacent to $\angle BCA$ and forms a straight line. Therefore, $\angle BCD$ and $\angle BCA$ are supplementary. The sum of supplementary angles totals 180°.

So      $x + y = 180°$

$$x + 50 = 180°$$

$$x = 130°$$

The correct choice is (D).

## • PROBLEM 1-68

Billy walked home from school 7 blocks east, 5 blocks north, 1 block west, and 3 blocks north again. How many blocks, in a straight line, is Billy's home from school?

(A) 5                (D) 16

(B) 10              (E) 20

(C) 15

## SOLUTION:

**(B)** It is easier to visualize the straight line to Billy's house from school if we draw a diagram.

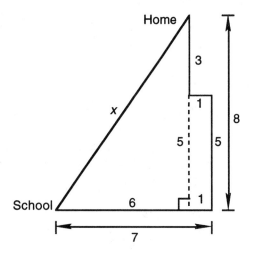

We can see that the shortest distance between Billy's home and school is the hypotenuse of a right triangle.

Since Billy walked 7 blocks east and 1 block west, he is 6 blocks east of school. Since he walked 5 blocks north and 3 blocks north, Billy is 8 blocks north of school.

Using the Pythagorean Theorem on the triangle that is formed, we find that the square of the hypotenuse equals the sum of the squares of the other two sides.

$$a^2 + b^2 = c^2$$

$$(6)^2 + (8)^2 = c^2$$

$$36 + 64 = c^2$$

$$100 = c^2$$

$$10 = c$$

(B) is the correct answer.

## • PROBLEM 1-69

The U.S. soccer team won 70% of the games they played. If they played a total of 20 games, how many games did they lose?

(A)  6                          (D)  12

(B)  8                          (E)  14

(C)  10

## SOLUTION:

**(A)**    If the U.S. soccer team won 70% of the games they played, then they lost 30% of the games. Since they played a total of 20 games, we need to convert 30% to a decimal and multiply to find the number lost.

$$30\% = .30$$

$$30\% \text{ of } 20 = .30 \, (20) = 6$$

The correct choice is (A).

## • PROBLEM 1-70

Find the area of a sector of a circle with a radius of 16 cm and a 45° arc?

(A)  6.24 cm$^2$                (D)  803.84 cm$^2$

(B)  100.48 cm$^2$              (E)  2,260.8 cm$^2$

(C)  720 cm$^2$

## SOLUTION:

**(B)**

$$\text{Radius} = 16 \text{ cm}$$

$$\text{Arc} = 45°$$

$$\pi = 3.14$$

The area of a sector equals

$$\left(\frac{\text{degrees of given arc}}{360}\right)\pi r^2.$$

Substituting the given values for the radius and arc, we have

$$\text{Area} = \left(\frac{45}{360}\right)3.14\,(16)^2.$$

Simplifying we get

$$\text{Area} = \frac{1}{8}(3.14)\,(256)$$

$$= 32(3.14)$$

$$= 100.48 \text{ cm}^2$$

The correct answer is (B).

### • PROBLEM 1–71

At a certain restaurant the cost of 3 sandwiches, 7 cups of coffee, and 4 pieces of pie is $10.20, while the cost of 4 sandwiches, 8 cups of coffee, and 5 pieces of pie is $12.25. What is the cost of a luncheon consisting of one sandwich, one cup of coffee, and one piece of pie?

(A) $2.00

(D) $2.15

(B) $2.05

(E) $2.25

(C) $2.10

## SOLUTION:

(B)    Let $x$ = cost of one sandwich

$y$ = cost of one cup of coffee

$z$ = cost of one piece of pie

So 3 sandwiches, 7 cups of coffee, and 4 pieces of pie equals $10.20 can be written as

$$3x + 7y + 4z = 10.20.$$

Similarly, 4 sandwiches, 8 cups of coffee, and 5 pieces of pie equal $12.25 can be written as

$4x + 8y + 5z = 12.25.$

If we subtract the two equations, we find that $x + y + z = 2.05$.

$$4x + 8y + 5z = 12.25$$
$$3x + 7y + 4z = 10.20$$
$$\overline{\phantom{3}x + \phantom{7}y + \phantom{4}z = \phantom{1}2.05}$$

This means that the cost of one sandwich, one cup of coffee, and one piece of pie is $2.05.

The correct choice is (B).

## • PROBLEM 1–72

An airplane travels 1,800 miles in 3 hours flying with the wind. On the return trip, flying against the wind it takes 4 hours to travel 2,000 miles. Find the rate of the plane in still air.

(A)  425 mph

(B)  500 mph

(C)  542 mph

(D)  550 mph

(E)  600 mph

## SOLUTION:

**(D)**    The actual speed of the plane in still air is equal to the speed of the plane minus the wind speed when going with the wind.

Likewise, the actual speed of the plane in still air is equal to the speed of the plane plus wind speed when going against the wind.

Using the formula distance = (rate) (time),

we see that

$$\text{Rate} = \frac{\text{Distance}}{\text{time}}.$$

The speed (or rate) of the airplane with the wind is

$$\text{Rate} = \frac{1{,}800 \text{ miles}}{3 \text{ hrs}} = 600 \text{ miles/hr}.$$

So 600 mi/hr – wind speed = actual rate of airplane flying with the wind.

On the return trip, the rate of the plane is found as follows:

$$\text{Rate} = \frac{2,000 \text{ miles}}{4 \text{ hrs}} = 500 \text{ miles/hr}$$

500 miles/hr + wind speed = actual rate of airplane flying against the wind.

Travelling with the wind produces a positive wind speed while traveling against the wind produces a negative wind speed.

Let    $w$ = wind speed

       $x$ = actual speed of plane in still air

       $x$ = 600 miles/hr − $w$           (with the wind)

       $x$ = 500 miles/hr + $w$          (against the wind)

Setting these two equations equal, we get

    600 miles/hr − $w$ = 500 miles/hr − ($-w$)

    600 miles/hr − $w$ = 500 miles/hr + $w$

Solving for $w$,

   $2w$ = 100 miles/hr

   $w$ = 50 miles/hr

Substituting $w$ = 50 into either initial equation gives us a value for $x$.

   $x$ = 600 miles/hr − 50 miles/hr

   $x$ = 550 miles/hr

The correct answer is (D).

## • PROBLEM 1-73

If a triangle of base 6 units has the same area as a circle of radius 6 units, what is the altitude of the triangle?

(A) $\pi$                     (D) $12\pi$

(B) $3\pi$                (E) $36\pi$

(C) $6\pi$

## SOLUTION:

**(D)**     The formula for the area of a triangle is

$$A = \frac{1}{2} \, bh$$

where $b$ denotes the base and $h$ denotes the altitude.

The formula for the area of a circle is

$$A = \pi r^2$$

where $\pi = 3.14$ and $r$ denotes the radius.

Since $b = 6$, then the area of the triangle is

$$A = \frac{1}{2}(6)\, h = 3h.$$

We are given that the radius is 6. Substituting this value for $r$ in the formula for the area of a circle, we obtain

$$A = \pi \, (6)^2 = 36\pi.$$

Since the area is the same for both figures, we can say that $3h = 36\pi$.

Solving for $h$, we find that

$$h = \frac{36\pi}{3} = 12\pi.$$

Therefore, the altitude of the triangle is $12\pi$.

The correct choice is (D).

### • PROBLEM 1–74

How many different segments are determined by 6 points on a line?

(A)  12                (D)  21

(B)  15                (E)  30

(C)  17

## SOLUTION:

**(B)**

A      B      C      D      E      F

Label the points *A, B, C, D, E,* and *F* as shown.

Then the line segments are

$$\overline{AB}$$

$$\overline{AC} \quad \overline{BC}$$

$$\overline{AD} \quad \overline{BD} \quad \overline{CD}$$

$$\overline{AE} \quad \overline{BE} \quad \overline{CE} \quad \overline{DE}$$

$$\underline{\overline{AF} \quad \overline{BF} \quad \overline{CF} \quad \overline{DF} \quad \overline{EF}}$$

$$5 \quad\quad 4 \quad\quad 3 \quad\quad 2 \quad\quad 1$$

There are 15 segments.

The correct choice is (B).

### • PROBLEM 1–75

It takes 15 apples to make 4 pies. How many pies can be made from 20 apples?

(A) 4

(B) $4\dfrac{1}{3}$

(C) $4\dfrac{2}{3}$

(D) 5

(E) $5\dfrac{1}{3}$

## SOLUTION:

**(E)**    The ratio of apples to pies is $\dfrac{15}{4}$.

We need to find how many pies we can make from 20 apples. So, we need to set up a proportion.

$$\frac{15}{4} = \frac{20}{x}$$

We solve for *x* by cross multiplying.

$$15x = 4\,(20)$$

$$15x = 80$$

$$x = \frac{80}{15}$$

$$x = 5\frac{1}{3}$$

The correct choice is (E).

**• PROBLEM 1-76**

Emile receives a flat weekly salary of $240 plus 12% commission of the total volume of all sales he makes. What must his dollar volume be in a week if he is to make a total weekly salary of $540?

(A)  $2,800          (D)  $2,500

(B)  $3,600          (E)  $2,000

(C)  $6,400

## SOLUTION:

**(D)**      Let $x$ = total value of all sales

$12\% \, x = .12x$ = Emile's commission

$240 + .12x$ = Emile's weekly salary

If Emile receives $540 one week, then

$$240 + .12x = 540$$

$$.12x = 300$$

$$x = \frac{\$300}{.12} = \$2,500$$

The correct choice is (D).

## • PROBLEM 1-77

The area of $\triangle ADE$, an equilateral triangle, is 12 square units. If $B$ is the midpoint of $\overline{AD}$ and $C$ is the midpoint of $\overline{AE}$, what is the area of $\triangle ABC$?

(A)  2 square units

(D)  4 square units

(B)  3 square units

(E)  6 square units

(C)  $3^1/_2$ square units

## SOLUTION:

**(B)**  Let $F$ be the midpoint of $\overline{DE}$. The four small triangles *ABC*, *CEF*, *BFD*, and *BCF* are then congruent. This means they have the same size and the same shape.

Hence, each small triangle is $^1/_4$ the area of the larger triangle $\triangle ADE$.

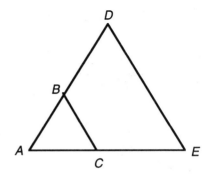

$$\text{area } \triangle ABC = \frac{1}{4} \text{ area } \triangle ADE$$

Therefore, the area of

$$\triangle ABC = \frac{1}{4}(12) = 3.$$

The correct choice is (B).

## • PROBLEM 1-78

If $6x + 12 = 5$, then the value of $(x + 2)$ is

(A)  $-\dfrac{19}{6}$.

(D)  $3\dfrac{1}{6}$.

(B)  $-1\dfrac{1}{6}$.

(E)  $1\dfrac{1}{6}$.

(C)  $\dfrac{5}{6}$.

## SOLUTION:

**(C)**    We must first solve for $x$.

$$6x + 12 = 5$$

$$6x = -7$$

$$x = -\frac{7}{6}$$

We want the value of $(x + 2)$. Since $x = -7/6$, we substitute this value for $x$ in the equation.

$$(x + 2) = \left(-\frac{7}{6}\right) + 2$$

$$(x + 2) = \left(-\frac{7}{6}\right) + \frac{12}{6}$$

$$x + 2 = \frac{5}{6}$$

Therefore, the correct answer is (C).

## • PROBLEM 1-79

What is the value of $x$ in the equation $\sqrt{5x - 4} - 5 = -1$?

(A)  2                             (D)  4

(B)  5                             (E)  $-4$

(C)  No value

## SOLUTION:

**(D)**

$$\sqrt{5x - 4} - 5 = -1$$

The best approach to this problem would be to isolate the radical. This is accomplished by adding 5 to both sides.

$$\sqrt{5x - 4} - 5 + 5 = -1 + 5$$

$$\sqrt{5x - 4} = 4$$

We can now square both sides.

$$(\sqrt{5x - 4})^2 = 4^2$$

$$5x - 4 = 16$$

Solving for $x$ we obtain

$$5x - 4 + 4 = 16 + 4$$

$$5x = 20$$

$$x = \frac{20}{5}$$

$$x = 4$$

The correct answer is (D).

## • PROBLEM 1-80

If $R$, $S$, and $Q$ can wallpaper a house in 8 hours and $R$ and $S$ can do it in 12 hours, how long will it take $Q$ alone to wallpaper the house?

(A)  12 hours

(D)  20 hours

(B)  24 hours

(E)  28 hours

(C)  8 hours

## SOLUTION:

**(B)**    Let $x$ = the number of hours it takes $Q$ to wallpaper the house by himself.

In 1 hour $Q$ will wallpaper $1/x$ of the house.

Since it takes $R$, $S$, and $Q$ eight hours to do the job if they work together, it follows that $R$, $S$, and $Q$ complete $1/8$ of the house in 1 hour.

If $R$ and $S$ work together, they complete the job in 12 hours. Hence, in 1 hour they can complete $1/12$ of the house.

If all three of them work together, they can wallpaper $(1/12 + 1/x)$ of the house in 1 hour. But we know that all three of them wallpaper $1/8$ of the house in 1 hour. Thus,

$$\frac{1}{x} + \frac{1}{12} = \frac{1}{8}$$

Solving this equation for $x$ yields the number of hours it takes $Q$ to wallpaper the house by himself.

$$\frac{1}{x} + \frac{1}{12} = \frac{1}{8}$$

$$\frac{1}{x} = \frac{1}{8} - \frac{1}{12}$$

We must simplify $1/8 - 1/12$ by find the lowest common denominator.

$$\frac{1}{8} = \frac{3}{24}$$

$$\frac{1}{12} = \frac{2}{24}$$

Therefore,

$$\frac{1}{x} = \frac{3}{24} - \frac{2}{24}$$

$$\frac{1}{x} = \frac{1}{24}$$

Cross multiplying we find that $x = 24$.

The correct choice is (B).

## • PROBLEM 1-81

If $a$ and $b$ are odd integers, which of the following must be an even integer?

I. $\dfrac{a+b}{2}$

II. $ab - 1$

III. $\dfrac{ab+1}{2}$

(A) I only.

(D) II and III only.

(B) II only.

(E) I, II, and III.

(C) I and II only.

## SOLUTION:

**(B)**   Let $a = 2n + 1$

$b = 2m + 1$

Take each case individually.

I.   $$\frac{a+b}{2} = \frac{(2n+1)+(2m+1)}{2}$$

$$= \frac{2n+2m+2}{2}$$

$$= \frac{2(n+m+1)}{2}$$

$$= n+m+1$$

This is not necessarily even.

II.   $ab - 1 = (2n + 1)(2m + 1) - 1$

$$= (4mn + 2n + 2m + 1) - 1$$

Combining the numerical terms and factoring out 2, we get

$ab - 1 = 2(2mn + m + n).$

This will always be divisible by 2.

III.   $$\frac{ab+1}{2} = \frac{(2n+1)+(2m+1)+1}{2}$$

$$= \frac{4mn + 2n + 2m + 1 + 1}{2}$$

$$= 2nm + m + m + 1$$

$$= 2nm + 2m + 1$$

$$= 2(nm + m) + 1$$

This is always odd.

Hence, only II is even.

The correct choice is (B).

**• PROBLEM 1-82**

Which of the following equations can be used to find a number $n$, such that if you multiply it by 3 and subtract 2, the result is 5 times as great as if you divide the number by 3 and add 2?

(A) $3n - 2 = 5 + \dfrac{n}{3} + 2$

(D) $5(3n - 2) = \dfrac{n}{3} + 2$

(B) $3n - 2 = 5\left(\dfrac{n}{3} + 2\right)$

(E) $5n - 2 = \dfrac{n}{3} + 2$

(C) $3n - 2 = \dfrac{5n}{3} + 2$

## SOLUTION:

**(B)** The best approach to this problem is to follow the directions given in sequence.

Let $\qquad n$ = the number

$\qquad 3n$ = three times the number

$\qquad 3n - 2$ = three times the number minus 2.

$3n - 2$ is five times as great as $\dfrac{n}{3}$ + 2.

$$3n - 2 = 5\left(\dfrac{n}{3} + 2\right)$$

The correct answer is (B).

**• PROBLEM 1-83**

If it takes $s$ sacks of grain to feed $c$ chickens, how many sacks of grain are needed to feed $k$ chickens?

(A) $\dfrac{ck}{s}$

(D) $\dfrac{c}{sk}$

(B) $\dfrac{k}{cs}$

(E) $\dfrac{sk}{c}$

(C) $\dfrac{cs}{k}$

## SOLUTION:

**(E)**    Let $s$ = number of sacks of grain

$c$ = number of chickens

$\dfrac{s}{c}$ = number of sacks of grain needed to feed one chicken.

If we have $k$ chickens, we will need to multiply the number of sacks of grain needed to feed one chicken by $k$. Therefore,

$$\frac{s}{c} \times k = \frac{sk}{c}$$

The answer is (E).

### • PROBLEM 1–84

In the five-pointed star shown, what is the sum of the measures of angles $A$, $B$, $C$, $D$, and $E$?

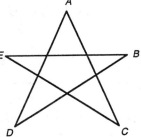

(A)  108°

(B)  72°

(C)  36°

(D)  150°

(E)  180°

## SOLUTION:

**(E)**    Let $m\angle A$ represent the measure of angle $A$. Though there are several ways to attack this question, one way is to recall that the sum of the measures of the three interior angles of a triangle is equal to 180°, and the measure of an exterior angle of a triangle is equal to the sum of the measures of the two nonadjacent interior angles of the triangle.

We can now start by considering triangle $ACL$. Of course,

$$m\angle A + m\angle C + m\angle 1 = 180°. \tag{1}$$

But $\angle 1$ is an exterior angle to triangle $LEF$, thus,

$$m\angle 1 = m\angle E + m\angle 2.$$

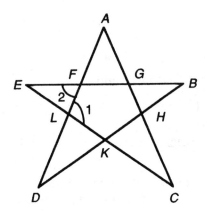

Substituting this in equation (1) yields,

$$m\angle A + m\angle C + m\angle E + m\angle 2 = 180°. \tag{2}$$

However, $\angle 2$ is an exterior angle to triangle *FBD*, thus,

$$m\angle 2 = m\angle B + m\angle D.$$

Substituting this result in equation (2) yields,

$$m\angle A + m\angle C + m\angle E + m\angle B + \mu\angle D = 180°.$$

Thus, the sum of the measures of angles *A, B, C, D,* and *E* is equal to 180°.

## • PROBLEM 1–85

Mary drove from Jamestown to Aberdeen, a distance of 100 miles at an average speed of 40 miles per hour. She made the return trip at an average speed of 60 miles per hour. What was her average speed for the round trip?

(A)  48 mph

(B)  49 mph

(C)  50 mph

(D)  51 mph

(E)  52 mph

### SOLUTION:

**(A)**

$$\text{Distance} = \text{rate} \times \text{time}$$

$$\text{Time} = \frac{\text{Distance}}{\text{rate}}$$

Going to Aberdeen from Jamestown took

$$\frac{100 \text{ miles}}{40 \text{ miles/hr}} = \frac{5}{2} \text{ hrs.}$$

The return trip to Jamestown took

$$\frac{100 \text{ miles}}{60 \text{ miles/hr}} = \frac{5}{3} \text{ hrs.}$$

Therefore, the total time for the round trip was

$$\frac{5}{2} + \frac{5}{3} = \text{total time.}$$

In order to add these fractions, we need to find a common denominator.

$$\frac{5}{2} = \frac{15}{6}$$

$$\frac{5}{3} = \frac{10}{6}$$

$$\frac{15}{6} + \frac{10}{6} = \frac{25}{6} \text{ hours}$$

The average speed or rate for the total trip is found by dividing the total distance travelled by the total time.

$$\frac{\text{total distance (round trip)}}{\text{total time}} = \frac{100 + 100}{\left(\frac{25}{6}\right)}$$

$$= \frac{200}{\frac{25}{6}}$$

$$200 \times \frac{6}{25} = 48 \text{ miles/hr}$$

The correct answer is (A).

## • PROBLEM 1–86

Which of the following statements are true, if

$$x + y + z = 10$$

$$y \geq 5$$

$$4 \geq z \geq 3$$

I.   $x < z$

II.  $x > y$

III. $x + z \leq y$

(A)  I only.

(B)  II only.

(C)  III only.

(D)  I and III.

(E)  I, II, and III.

## SOLUTION:

**(D)**   $x + y + z = 10$

$y \geq 5$

$4 \geq z \geq 3$

Solving for $x$ we obtain

$x = 10 - y - z.$

If we use the smallest values for $y$ and $z$ ($y = 5$ and $z = 3$), we obtain the largest value for $x$.

$$x = 10 - y - z$$

$$= 10 - 5 - 3$$

$$= 2$$

This implies that $x \leq 2$. Then $x < z$ and $x < y$. Consider again

$$x = 10 - y - z$$

$$x + z = 10 - y$$

Now rearrange this expression to analyze proposition III.

If     $y = 5$ (the smallest value),

$x + z = 5.$

If     $y > 5$, then $x + z < 5$.

Therefore, $x + z \le y$ making both I and III correct.

The correct choice is (D).

## • PROBLEM 1-87

Which of the following equations can be used to find a woman's present age if she is now 6 times as old as her son and next year her age will be equal to the square of her son's age?

(A)  $6w + 1 = w^2 + 1$      (D)  $6w + 1 = (w + 1)^2$

(B)  $6(w + 1) = w^2 + 1$      (E)  $w + 6 = (w + 1)^2$

(C)  $6(w + 1) = (w + 1)^2$

## SOLUTION:

**(D)**     Let

$w$ = age of the son now

$w + 1$ = age of the son next year

$6w$ = age of the mother now

$6w + 1$ = age of the mother next year

We are given that next year the age of the mother $(6w + 1)$ will equal the square of her son's age $(w + 1)^2$.

So:

$6w + 1 = (w + 1)^2.$

The correct choice is (D).

### • PROBLEM 1-88

If the length of a rectangle is increased by 30% and the width is decreased by 20%, then the area is increased by

(A) 10%.    (D) 20%.

(B) 5%.    (E) 25%.

(C) 4%.

## SOLUTION:

**(C)**    Let

x = length of the rectangle

y = width of the rectangle

If the length is increased by 30%, we can represent the new length by

$x + .30x = 1.3x.$

If the width is decreased by 20%, we can represent the new width by

$y - .20y = .80y.$

Since the area of a rectangle is found by multiplying length times width, the new area is found as follows:

new area = $(1.3x)(.80y)$

Simplifying we get

new area = $1.04xy.$

The area of the old rectangle is $xy$.

area new rectangle − area old rectangle =

$1.04\,xy - 1.0\,xy = .04\,xy$

We can convert the coefficient .04 to a percent by multiplying by 100.

$.04 \times 100\% = 4\%$

The correct choice is (C).

## • PROBLEM 1–89

What is the smallest positive number that leaves a remainder of 2 when the number is divided by 3, 4, or 5?

(A) 22          (D) 122

(B) 42          (E) 182

(C) 62

### SOLUTION:

**(C)** We must first find the least common multiple (LCM) of 3, 4, and 5. This is obtained simply by multiplying $3 \times 4 \times 5$.

$3 \times 4 \times 5 = 60$

60 is the LCM.

60 is divisible by 3.

60 is divisible by 4.

60 is divisible by 5.

In order to guarantee that the remainder in each case upon division by 3, 4, or 5 is 2, we simply add 2 to 60 to get 62.

The correct answer is (C).

## • PROBLEM 1–90

If $\dfrac{a}{x} - \dfrac{b}{y} = c$ and $xy = \dfrac{1}{c}$, then $bx =$

(A) $1 - ay$.          (D) $ay - 1$.

(B) $ay$.          (E) $2ay$.

(C) $ay + 1$.

### SOLUTION:

**(D)** $\dfrac{a}{x} - \dfrac{b}{y} = c$ and $xy = \dfrac{1}{c}$

The first step would be to combine the terms on the left side of the equations over a common denominator.

$$\frac{y}{y} \times \frac{a}{x} - \frac{x}{x} \times \frac{b}{y} = c$$

$$\frac{ay - bx}{xy} = c$$

We can now simplify the expression and solve for $bx$.

$$\frac{ay - bx}{xy} = c$$

Multiplying both sides of the equation by $xy$, we obtain

$$\frac{xy(ay - bx)}{xy} = c\,xy$$

$$ay - bx = c\,xy$$

Subtracting $ay$ from both sides and multiplying by $-1$, we obtain

$$bx = ay - c\,xy$$

Since we know that $xy = \frac{1}{c}$, we may substitute this value in the equation.

$$bx = ay - c\left(\frac{1}{c}\right)$$

This can be further simplified to give us the following:

$$bx = ay - 1$$

The correct choice is therefore (D).

## • PROBLEM 1–91

Joe and Jim together have 14 marbles. Jim and Tim together have 10 marbles. Joe and Tim together have 12 marbles. What is the maximum number of marbles that any of these may have?

(A) 7

(B) 8

(C) 9

(D) 10

(E) 11

## SOLUTION:

**(B)** Let

$x$ = the number of marbles Joe owns

$y$ = the number of marbles Jim owns

$z$ = the number of marbles Tim owns

It is given that

$$x + y = 14 \tag{1}$$

$$y + z = 10 \tag{2}$$

$$x + z = 12 \tag{3}$$

Solving for $y$ in equation (2) we find

$$y + z = 10$$

$$y = 10 - z$$

Solving for $x$ in equation (3) we find

$$x + z = 12$$

$$x = 12 - z$$

$x$ and $y$ can now be written in terms of a common variable, $z$. We can now substitute these terms in equation (1) to obtain

$$x + y = 14 \tag{1}$$

$$(12 - z) + (10 - z) = 14$$

$$22 - 2z = 14$$

$$22 = 14 + 2z$$

$$22 - 14 = 2z$$

$$8 = 2z$$

$$4 = z$$

$\therefore$ We now know that Tim owns 4 marbles.

Since $x = 12 - z$, we can substitute 4 for the value of $z$ to obtain the number of marbles owned by Joe.

$$x = 12 - z = 12 - 4 = 8$$

$\therefore$ Joe owns 8 marbles.

Likewise, since $y = 10 - z$, we can find the number of marbles owned by Jim.

$$y = 10 - z = 10 - 4 = 6$$

$\therefore$ Jim owns 6 marbles.

Joe's marbles, 8, is the maximum number of marbles anyone can have.

The correct choice is (B).

## • PROBLEM 1-92

In the Klysler Auto Factory, robots assemble cars. If 3 robots assemble 17 cars in 10 minutes, how many cars can 14 robots assemble in 45 minutes if all robots work at the same rate all the time?

(A) 357

(D) 150

(B) 340

(E) 272

(C) 705

## SOLUTION:

**(A)** If 3 robots can assemble 17 cars in 10 minutes, then 3 robots can assemble $^{17}/_{10}$ cars in 1 minute. To find the number of cars that 1 robot can assemble in 1 minute, we divide $^{17}/_{10}$ by 3.

One robot assembles

$$\frac{1}{3}\left(\frac{17}{10}\right) \text{ or } \frac{17}{30}$$

of a car in 1 minute.

Similarly, if 14 robots assemble $x$ cars in 45 minutes, then the 14 robots assemble $^{x}/_{45}$ cars in 1 minute.

Thus, 1 robot assembles

$$\frac{1}{14}\left(\frac{x}{45}\right) \text{ or } \frac{x}{14(45)}$$

of a car in 1 minute.

Since the robots all work at the same rate, we can say that

$$\frac{x}{14(45)} = \frac{17}{30}$$

$$\frac{x}{630} = \frac{17}{30}$$

Cross multiplying we get

$$30x = 17(630)$$

$$30x = 10,710$$

Dividing both sides of the equation by 30, we get

$$\frac{30x}{30} = \frac{10,710}{30}$$

$$x = 357$$

The correct choice is (A).

## • PROBLEM 1-93

A postal truck leaves its station and heads for Chicago, averaging 40 mph. An error in the mailing schedule is spotted and 24 minutes after the truck leaves, a car is sent to overtake the truck. If the car averages 50 mph, how long will it take to catch the postal truck?

(A)  2.6 hours           (D)  1.5 hours

(B)  3 hours             (E)  1.6 hours

(C)  2 hours

## SOLUTION:

**(E)**     Let $t$ = time in hours it takes the car to catch up with the postal truck

$$\left(t + \frac{24}{60}\right) = \text{time of travel of the truck}$$

$(t + .4)$ = time of travel in hours of the truck

For the truck:

Distance = rate × time

$$D = 40\,(t + .4)$$

$$D = 40t + 16$$

For the car:

Distance = rate × time

$$D = 50 \times t$$

$$D = 50t$$

Since the distance travelled by the truck and the car are equal, we can set these two equations equal to each other to get

$$50t = 40t + 16.$$

Solving for $t$ we get

$$50t - 40t = 16$$

$$10t = 16$$

$$t = 1.6$$

Therefore, it takes the car 1.6 hrs to catch up with the postal truck.

The correct choice is (E).

## • PROBLEM 1-94

A table tennis tournament is to be round-robin; that is, each player plays one match against every other player. The winner of the tournament is determined by the best scores in the matches. How many matches will be played if 5 people enter the tournament?

(A)  10                          (D)  105

(B)  15                          (E)  120

(C)  20

## SOLUTION:

**(A)**      Label the players 1 through 5. Player 1 will have one match with each of the remaining 4 players. Player 2 (who has already played Player 1) will have matches with the remaining three players and so on. The total number of matches will be

4 + 3 + 2 + 1 = 10.

The choice choice is (A).

## • PROBLEM 1-95

The most economical price among the following prices is

(A)  10 oz. for 16¢.                    (D)  20 oz. for 34¢.

(B)  2 oz. for 3¢.                       (E)  8 oz. for 13¢.

(C)  4 oz. for 7¢.

## SOLUTION:

**(B)**    To find the most economical price we must find out what 1 oz. would cost. To do this we simply divide the price by the number of ounces.

(A)     10 oz. for 16¢ gives us 1 oz. for $\dfrac{16}{10} = 1.6$¢

(B)     2 oz. for 3¢ gives us 1 oz. for $\dfrac{3}{2} = 1.5$¢

(C)     4 oz. for 7¢ gives us 1 oz. for $\dfrac{7}{4} = 1.75$¢

(D)     20 oz. for 34¢ gives us 1 oz. for $\dfrac{34}{20} = 1.70$¢

(E)     8 oz. for 13¢ gives us 1 oz. for $\dfrac{13}{8} = 1.63$¢

The most economical value is therefore (B)

## • PROBLEM 1-96

Pipe 1 can fill a tank in 3 hours. Pipe 2 can fill the same tank in 5 hours. To the nearest hour, how long would it take both pipes working together to fill the tank?

(A) 1                 (D) 4

(B) 2                 (E) 5

(C) 3

## SOLUTION:

**(B)** Let $x$ = the number of hours it would take both pipes working together to fill the tank.

Pipe 1 will fill $1/3$ of the tank in 1 hr.

Pipe 2 will fill $1/5$ of the tank in 1 hr.

In $x$ hours Pipe 1 will fill $x/3$ of the tank and Pipe 2 will fill $x/5$ tank.

Adding these two amounts together we have a full tank

$$\frac{x}{3} + \frac{x}{5} = 1$$

$$\frac{5x}{5 \times 3} + \frac{3x}{3 \times 5} = 1$$

$$\frac{8x}{15} = 1$$

$$8x = 15$$

$$x = \frac{15}{8} = 1\frac{7}{8} \text{ hrs}$$

To the nearest hour this is 2.

The correct choice is (B).

## • PROBLEM 1-97

A square is cut from a circle as shown in the diagram. If the radius of the circle is 6, what is the total area of the shaded regions?

(A) $9\pi - 18$

(B) $12\pi - 18$

(C) $12\pi - 72$

(D) $36\pi - 36$

(E) $36\pi - 72$

## SOLUTION:

**(E)** Area of shaded region = area of circle − area of square

The area of the circle is $\pi(\text{radius})^2 = \pi(6)^2 = 36\pi$

The area of a square is $(\text{side})^2$.

Consider $1/4$ of the square where the side of the square is equal to the hypotenuse of the triangle. The legs of the triangle are then radii of the circle.

The area of the triangle is

$$\frac{1}{2}bh = \frac{1}{2}(6)(6) = 18.$$

Since we have four such triangles making up the square, the area of the square is $4 \times 18 = 72$.

The area of the shaded region is therefore $36\pi - 72$.

The correct choice is (E).

• PROBLEM 1–98

For non-zero numbers, $p$, $q$, $r$, and $s$, $\dfrac{p}{q} = \dfrac{r}{s}$. Which of the following statements is true?

(A) $\dfrac{p+q}{p} = \dfrac{r+s}{r}$

(D) $\dfrac{p-q}{q} = \dfrac{s-r}{r}$

(B) $\dfrac{p}{r} = \dfrac{q}{r}$

(E) $\dfrac{q}{p-q} = \dfrac{r}{s-r}$

(C) $\dfrac{r}{s} = \dfrac{q}{r}$

## SOLUTION:

**(A)**   $p$, $q$, $r$, and $s$ are non-zero numbers.

$$\frac{p}{q} = \frac{r}{s}.$$

If we cross multiply, we see that

$$ps = qr. \qquad (1)$$

In answer choice (A), we see that

$$\frac{p+q}{p} = \frac{r+s}{r}.$$

Cross multiplying will give us

$$r(p+q) = p(r+s).$$

Using the distribution law, we get

$$rp + rq = pr + ps.$$

Since $rp$ and $pr$ are equal, we can rewrite the equation as follows:

$$rp + rq = rp + ps$$

Subtracting $rp$ from both sides, we get

$$rp + rq - rp = rp + ps - rp$$

$$rq = ps$$

Since $rq = qr$, then $qr = ps$. Thus, choice (A), when rearranged, is the same as equation (1), which is known to be true.

Therefore, choice (A) is correct.

## • PROBLEM 1-99

Each of the integers $h$, $m$, and $n$ is divisible by 3. Which of the following integers is *always* divisible by 9?

I.    $hm$

II.   $h + m$

III.  $h + m + n$

(A) I only.                    (D) II and III only.

(B) II only.                   (E) I, II, and III.

(C) III only.

## SOLUTION:

**(A)**    $h$, $m$, and $n$ are divisible by 3.

Let    $h = 3x$

        $m = 3y$

        $n = 3z$

I.    $hm = (3x)(3y) = 9xy$

This is divisible by 9.

II.    $h + m = 3x + 3y = 3(x + y)$

This is not necessarily divisible by 9.

III.    $h + m + n = 3x + 3y + 3z$

            $= 3(x + y + z)$

This is not necessarily divisible by 9.

So, the correct choice is (A).

### • PROBLEM 1–100

In the figure, if $\angle AOB = 60°$ and $O$ is the center of the circle with radius = 6, then the area of the shaded region is

(A) $6\pi$.

(B) $6\pi - 2\sqrt{3}$.

(C) $6\pi - 4\sqrt{3}$.

(D) $6\pi - 6\sqrt{3}$.

(E) $6\pi - 9\sqrt{3}$.

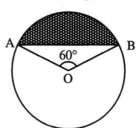

**SOLUTION:**

**(E)**    $\overline{OA}$ and $\overline{OB}$ are radii of the circle.

We are given that the radius is 6.

Since $\overline{OA} = \overline{OB}$, then $\angle OAB = \angle OBA$.

Since the sum of the angles of a triangle is equal to 180°, then

$$\angle OAB + \angle OBA + \angle AOB = 180°$$

Since $\angle OAB = \angle OBA$, we can substitute $\angle OAB$ in place of $\angle OBA$. We also know that $\angle AOB$ is equal to 60°. Therefore

$$\angle OAB + \angle OAB + 60° = 180°$$

$$2 \angle OAB + 60° = 180°$$

$$2 \angle OAB = 120°$$

$$\angle OAB = 60°$$

Since $\angle OAB = \angle OBA$, then we can say that $\angle OBA$ is also 60°. Therefore $\triangle AOB$ is equilateral and then $\overline{AB}$ must equal 6.

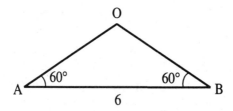

We must find the area of the triangle. The area of the triangle is

$1/2$ *bh.* We need to compute $h$ by using the Pythagorean Theorem.

$$a^2 + h^2 = c^2$$

$$3^2 + h^2 = 6^2$$

$$h^2 = 36 - 9 = 27$$

$$h = \sqrt{27} = \sqrt{9} \times \sqrt{3} = 3\sqrt{3}$$

Therefore, the area of triangle $AOB = 1/2$ (base) (height):

$$\frac{1}{2} \times 6 \times 3\sqrt{3} = 9\sqrt{3}$$

The area of sector $AOB$ equals

$$\left(\frac{60}{360}\right) \pi(r^2) = \frac{1}{6}\pi(6)^2$$

$$= \frac{1}{6}\pi \times 36$$

$$= 6\pi$$

Finally, we can compute the area of the shaded region as follows:

Area of shaded region = area of sector $AOB$ – area of triangle $AOB$.

Substituting our values for the area of sector $AOB$ and the area of triangle $AOB$, we obtain

Area of shaded region = $6\pi - 9\sqrt{3}$

The correct choice is (E).

# Chapter 2
# Quantitative Comparisons

# CHAPTER 2

# QUANTITATIVE COMPARISONS

In the Quantitative Comparison section of the SAT I you are asked to compare two quantities. Since you only have to compare quantities, the questions in this section usually take less time to solve than regular multiple-choice questions.

The two quantities are always presented in two columns, Column A and Column B.

| <u>Column A</u> | <u>Column B</u> |
|:---:|:---:|
| $2^3$ | $2^2$ |

Your job is to determine which quantity, if either, is greater. The directions tell you that there are four possible answer choices:

- If the quantity in Column A is greater than the quantity in Column B, the correct choice is (A).

- If the quantity in Column B is greater than the quantity in Column A, the correct choice is (B).

- If the two quantities are equal, the correct choice is (C).

- If the relationship cannot be determined, the correct choice is (D).

What would the correct choice be for the problem above? Since the quantity in Column A is equal to 8 and the quantity in Column B is equal to 4, and $8 > 4$, the correct choice is (A).

## ABOUT THE DIRECTIONS

Now that you have seen some of the directions, let's go over the rest of them. It's important to become familiar with the directions before the test so you don't waste time reading them during the test. You will also see the same reference information that appeared before the regular multiple-choice problems. Become familiar with this material before you take the SAT I. If you have to refer to any part of this page during the test, you are taking away valuable time needed to solve the problems.

Each of the following questions consist of two quantities, one in Column A and one in Column B. You are to compare the two quantities and on the answer sheet blacken space

(A) if the quantity in Column A is greater;

(B) if the quantity in Column B is greater;

(C) if the two quantities are equal;

(D) if the relationship cannot be determined from the information given.

**NOTES:**

1. In certain questions, information concerning one or both of the quantities to be compared is centered above the two columns.

2. In a given question, a symbol that appears in both columns represents the same thing in Column A as it does in Column B.

3. Letters such as $x$, $n$, and $k$ stand for real numbers.

**EXAMPLES:**

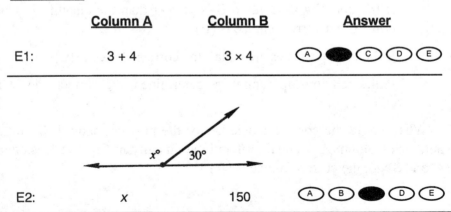

| | Column A | Column B | Answer |
|---|---|---|---|
| E1: | 3 + 4 | 3 × 4 | (A) ● (C) (D) (E) |
| E2: | $x$ | 150 | (A) (B) ● (D) (E) |

## ABOUT THE QUESTIONS

The following section will take you through the different types of Quantitative Comparison questions that you will encounter on the SAT I. Five main types exist and will be described here.

### Question Type 1: Calculations

This type of question will ask you to perform addition, subtraction, multiplication, or division. The question may ask you to perform one or more of these operations in order to compare the quantities in Columns A and B.

*PROBLEM*

| Column A | Column B |
| --- | --- |
| (13 + 44) (37 – 40) | (13 – 44) (37 – 40) |

*SOLUTION*

You could do the addition and subtraction in each column, but if you use your number knowledge, you can get the right answer in a matter of seconds! Try working with just signs. Remember that a smaller number minus a larger number yields a negative result so the answers to all the subtractions are negative. You'll get

$$(+) \, (–) \qquad \text{and} \qquad (–) \, (–)$$

which becomes

$$(–) \qquad \text{and} \qquad (+)$$

The correct choice is (B).

### Question Type 2: Conversions

This type of question may require you to convert the quantities given by asking you to find a percentage of a number, find a fraction of a number, find the area if given length and width, etc., in order to compare the quantities in Columns A and B.

*PROBLEM*

| Column A | Column B |
| --- | --- |
| 60% of $\frac{3}{4}$ | 75% of $\frac{3}{5}$ |

## SOLUTION

*Change the way the quantities look*; change the percents to fractions. You'll get

$$\left(\frac{3}{5}\right)\left(\frac{3}{4}\right) \qquad\qquad \left(\frac{3}{4}\right)\left(\frac{3}{5}\right)$$

You can now *compare parts*. The correct choice is (C).

## Question Type 3: Exponents and Roots

This type of question will ask you to determine a number's value, and then may ask you to use the value to perform an indicated operation in order to compare the quantities in Columns A and B.

### PROBLEM

| Column A | Column B |
|:---:|:---:|
| $\sqrt{.81}$ | $\sqrt[3]{.81}$ |

## SOLUTION

In this problem, $\sqrt{.81}$ is easy to find—it's 0.9. The problem is finding $\sqrt[3]{.81}$. There is no procedure to find $\sqrt[3]{.81}$ except estimating and multiplying. Since you are really concerned with seeing how $\sqrt[3]{.81}$ compares to 0.9 and not finding out exactly what it is, try 0.9 as an estimate. That means you have to cube 0.9. $0.9^3$ is .729. Since .729 is less than .81, the $\sqrt[3]{.81}$ must be greater than 0.9. The correct choice is (B).

## Question Type 4: Variables

This type of question will present you with variables of unknown quantities. You will be provided with additional information which must be applied in order to compare the quantities in Columns A and B.

### PROBLEM

| Column A | Column B |
|:---:|:---:|
| $y^* = y(y + 1)$ if $y$ is even | |
| $y^* = y(y - 1)$ if $y$ is odd | |
| $7^*$ | $6^*$ |

*SOLUTION*

Change the way the quantities look by applying the rules that appear between the two columns.

$$7* = (7)\,(7-1) \qquad\qquad 6* = 6(6+1)$$

$$7* = (7)\,(6) \qquad\qquad\qquad 6* = (6)\,(7)$$

The factors are identical in the two columns. The correct choice is (C).

## Question Type 5: Figures

This type of question will include a figure. You will be asked to use the figure provided to compare the quantities in Columns A and B. The figures may or may not be drawn to scale.

*PROBLEM*

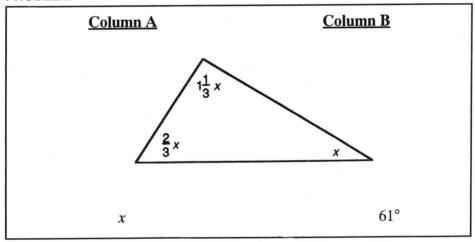

*SOLUTION*

The sum of the three angles of a triangle equals 180°. Therefore, we can add together the angles in the triangle above and set them equal to 180° to solve for $x$.

$$1\frac{1}{3}x + \frac{2}{3}x + x = 180$$

If we multiply this equation by three, we can eliminate the fractions.

$$3\,(1\frac{1}{3}x + \frac{2}{3}x + x = 180)$$

$$4x + 2x + 3x = 540$$

Simplifying the equation, we obtain

$$9x = 540$$

$$x = 60$$

Therefore, the correct choice is (B), since 60° is less than 61°.

As you attempt to answer Quantitative Comparison questions, it is not completely necessary that you be able to classify each question. However, you should be familiar with how to handle each question type. The remainder of this review will teach you how to answer the questions.

## ANSWERING QUANTITATIVE COMPARISON QUESTIONS

As you go through Quantitative Comparison questions, some seem to require a lot of calculations. On the other hand, there are some questions for which the answer seems immediately obvious. Often neither is the case. Excessive calculations can often be avoided and questions for which the answer seems obvious are often tricky.

The following steps will help guide you through answering Quantitative Comparison questions. These steps will cover strategies for every situation that you may encounter. Therefore, you may not need to go through every step for every question.

| STEP 1 | When you are presented with an addition or multiplication problem, you can often avoid these calculations just by comparing individual quantities. |

**PROBLEM**

| Column A | Column B |
|---|---|
| $\dfrac{1}{3} + \dfrac{1}{6} + \dfrac{1}{12}$ | $\dfrac{1}{4} + \dfrac{1}{8} + \dfrac{1}{16}$ |

**SOLUTION**

Both columns contains an addition problem. Don't bother to add! Compare the fractions in each column.

| Column A | | Column B |
|---|---|---|
| $\dfrac{1}{3}$ | $>$ | $\dfrac{1}{4}$ |

$$\frac{1}{6} \quad > \quad \frac{1}{8}$$

and $$\frac{1}{12} \quad > \quad \frac{1}{16}$$

For each fraction in Column B, there is a greater fraction in Column A. Therefore, the sum in Column A must be greater than the sum in Column B. The correct choice is (A).

**PROBLEM**

| Column A | Column B |
|---|---|
| The area of a rectangle with a base of 3 and a height of 4 | The area of a circle with a radius of 2 |

**SOLUTION**

You have to compare the area of a rectangle to the area of a circle. Your two columns will look like

| Area of rectangle | Area of circle |
|---|---|
| $(3)(4)$ | $\pi 2^2 = 4\pi$ |

Both columns contain a multiplication problem with positive numbers. Don't multiply! Compare the factors in each column.

$$3 \quad < \quad \pi$$

and $$4 \quad = \quad 4$$

The factors in Column B are greater than or equal to the factors in Column A so the product in Column B (the area of the circle) must be greater than the product in Column A (the area of the rectangle). The correct choice is (B).

| STEP 2 | Sometimes you are faced with problems in which comparing parts seems to apply but you can't compare the quantities the way they are presented in the problem. If you change the way the quantities look, you may be able to compare parts. |
|---|---|

**PROBLEM**

| Column A | Column B |
|---|---|
| 20% of 368 | $\dfrac{1}{4}$ of 368 |

**SOLUTION**

Both columns contain multiplication of positive quantities.

$$(20\%)\,(368) \qquad\qquad \left(\frac{1}{4}\right)(368)$$

This means that you can compare individual quantities to avoid multiplication. Both columns contain 368, but Column A contains a percent and Column B contains a fraction. You can either change the percent to a fraction or the fraction to a percent.

If you change 20% to a fraction, the columns will look like

$$\left(\frac{1}{5}\right)(368) \qquad\qquad \left(\frac{1}{4}\right)(368)$$

Now you can compare quantities.

$$\frac{1}{5} < \frac{1}{4}$$

$$368 = 368$$

or

If you change $^1/_4$ to a percent, the columns will look like

$$(20\%)\,(368) \qquad\qquad (25\%)\,(368)$$

Now you can compare quantities in the two columns.

$$20\% < 25\%$$

$$368 = 368$$

Using either approach, the product in Column B is greater than the product in Column A. The correct choice is **(B)**.

**PROBLEM**

| Column A | Column B |
|---|---|
| The area of a rectangle with length of 2.5 feet and width of 2 feet | The area of a rectangle with length of 27 inches and width of 23 inches |

**SOLUTION**

Since you multiply length by width to find the area of a rectangle, both columns contain multiplication of positive quantities.

$$(2.5)\,(2) \qquad\qquad (27)\,(23)$$

If you write the problem this way and forget about the units, you have:

$$2.5 < 27$$

$$2 < 23$$

The product in Column B seems to be greater than the product in Column A. But this is wrong! Column A contains feet and Column B contains inches. Make sure you change feet to inches or inches to feet before you make the comparisons!

Let's change the feet to inches and see what happens. 2.5 feet equals 30 inches and 2 feet equals 24 inches. Now all the quantities are in inches so you can make the comparisons. Your two columns look like

$$(30)\,(24) \qquad\qquad (27)\,(23)$$

Now compare.

$$30 > 27$$

$$24 > 23$$

The product in Column A is greater than the product in Column B. The correct choice is (A).

**PROBLEM**

| Column A | Column B |
|---|---|
| The number of ounces in 2 pints | The number of pints in in 2 gallons |

## SOLUTION

In this problem you must make sure to convert to the units specified in each column.

| 1 pint = 16 ounces | 1 gallon = 8 pints |
|---|---|
| 2 pints = 32 ounces | 2 gallons = 16 pints |

There are more *ounces* in 2 pints (Column A) than there are *pints* in 2 gallons (Column B). The correct choice is (A).

## PROBLEM

| Column A | Column B |
|---|---|
| (4.37) (125) | (437) (1.25) |

## SOLUTION

Each column contains a multiplication problem with positive numbers but comparing parts doesn't work since

| | Column A | | Column B |
|---|---|---|---|
| | 4.37 | < | 437 |
| but | 125 | > | 1.25 |

Even if you switch the numbers around, it still won't help.

| | 125 | < | 437 |
|---|---|---|---|
| but | 4.37 | > | 1.25 |

You may think that you are forced to multiply, but you don't have to if you notice that all the digits are the same. The only differences between the quantities in the two columns are where the decimal points are. You can move the decimal points to try to get some or all of the numbers to look the same and then make the comparisons. When you move the decimal points, you are really multiplying or dividing by powers of 10 and what you do in one column must be done in the other column.

You can move the decimal point in 4.37 to the right two places in Column A (you're really multiplying by 100). You must then move the decimal point in Column B two places to the right.

| Column A | Column B |
|---|---|
| (4.37) (125) | (437) (1.25) |

becomes

$$(437)\,(125) \qquad\qquad (437)\,(125)$$

Now you have identical factors in both columns. The correct choice is (C).

**PROBLEM**

| Column A | Column B |
|:---:|:---:|
| $\dfrac{(14)\,(15)}{(35)\,(8)}$ | 1 |

**SOLUTION**

Sometimes changing the way quantities look won't help you compare parts, but it will make the arithmetic a lot easier!

Column A contains multiplication and division of positive numbers. Factor and reduce and you will make the problem a lot easier to solve.

When you factor,

| Column A | Column B |
|:---:|:---:|
| $\dfrac{(14)\,(15)}{(35)\,(8)}$ | 1 |

becomes

| | |
|:---:|:---:|
| $\dfrac{(2)\,(7)\,(3)\,(5)}{(7)\,(5)\,(2)\,(4)}$ | 1 |

When you reduce, you get

| | |
|:---:|:---:|
| $\dfrac{3}{4}$ | 1 |

The quantity in Column B is greater than the quantity in Column A, so the correct choice is (B).

---

| STEP 3 | Make sure to look at the numbers presented. Numbers have important properties that can be used to save time. There are many properties that will be useful to know for the SAT I. We'll highlight two here. |

**PROBLEM**

| Column A | Column B |
|----------|----------|
| (17) (15) (13) | (18) (16) (14) (0) |

**SOLUTION**

You may be tempted to compare factors in this problem, but you don't have to if you remember to look at both columns before you do any work. Notice that 0 is a factor in Column B. The quantity in Column B is equal to 0, since 0 times any number is 0. The quantity in Column A is greater than 0, since all the factors are positive. The correct choice is (A).

**PROBLEM**

| Column A | Column B |
|----------|----------|
| $(2)^5 + (-2)^5$ | $(3)^5 + (-3)^5$ |

**SOLUTION**

When you first look at this problem, you probably want to compare parts. To do that, you might think that you have to find the values of each part, but you don't! Getting the values involves a lot of multiplication so there must be a faster way and there is. Analyze the numbers in each column.

| | |
|---|---|
| $(2)^5$ = some positive number | $(3)^5$ = some positive number |
| $(-2)^5$ = the same as $(2)^5$ but with a negative sign | $(-3)^5$ = the same as $(3)^5$ but with a negative sign |

The two numbers in each column are the same but opposite in sign. What does that mean when you add the two numbers in each column? The sum in each column is 0; therefore, the quantities in the two columns are equal. The correct choice is (C).

**PROBLEM**

| Column A | Column B |
|----------|----------|
| $\dfrac{(-1)^7 + (-1)^9}{(-1)^4 + (-1)^6}$ | 1 |

**SOLUTION**

When $-1$ is raised to an odd power, the result is $-1$. When $-1$ is raised to an even power, the result is + 1. Using that information, you get the following:

$$\frac{(-1)+(-1)}{(+1)+(+1)} \qquad\qquad 1$$

$$\frac{-2}{+2} \qquad\qquad 1$$

$$-1 \qquad\qquad 1$$

The correct choice is (B).

| STEP 4 | Try substituting values for the variables, in particular 0, 1, – 1, and fractions between 0 and – 1. Sometimes you can use values that appear in the problem. To get a better understanding of how this will help you solve problems, let's look at a few.

**PROBLEM**

| **Column A** | **Column B** |
|:---:|:---:|
| | $x$ is positive | |
| $2^x$ | $2^3$ |

**SOLUTION**

In order to compare the two quantities, you have to know the value of $x$. According to the information centered above the two columns, $x$ can be any positive number. You could try different values for $x$, but where do you begin? Column B gives you a starting point. What happens if $x = 3$? The quantities in the two columns would be equal. Now that you have tried one value for $x$ and established that the two quantities can be equal, try another value to see if you can make one quantity greater than the other. What if $x > 3$? Then the quantity in Column A would be greater than the quantity in Column B (you shouldn't have to do the arithmetic!). By making these two comparisons, you have determined that:

the quantity in Column A = the quantity in Column B

or

the quantity in Column A > the quantity in Column B

The correct choice is (D).

(It's unnecessary to check values for $x < 3$ since you have already determined that (D) is the correct choice.)

**PROBLEM**

| Column A | Column B |
| --- | --- |
| The average of $a$, $b$, and $c$ is 10. | |
| $a > b > c$ | |
| $a - b$ | $b - c$ |

**SOLUTION**

Try some values for $a$, $b$, and $c$. The choice of numbers is up to you, but keep it simple. If you have 3 numbers with an average of 10, then their sum is 30. This may help you choose numbers.

You could start with $a = 11$, $b = 10$, and $c = 9$. Then you would have:

$$a - b = 11 - 10 = 1 \qquad \text{and} \qquad b - c = 10 - 9 = 1$$

The quantities in the two columns can be equal. But can there be some other relationship? Try some more numbers.

What if $a = 12$, $b = 11$, and $c = 7$? Then you would have

$$a - b = 12 - 11 = 1 \qquad \text{and} \qquad b - c = 11 - 7 = 4$$

In this case, the quantity in Column B is greater than the quantity in Column A.

You have determined that

the quantity is Column A = the quantity in Column B

or

the quantity in Column A < the quantity in Column B.

The correct choice is (D).

**PROBLEM**

| Column A | Column B |
| --- | --- |
| $x$ is positive. | |
| $x \neq 1$ | |
| $x$ | $x^2$ |

**SOLUTION**

Once again you have to try different values for $x$. Keep the arithmetic as simple as possible.

If $x = 2$, then you get

| 2 | 4 |

If $x = 3$, then you get

| 3 | 9 |

The answer seems obvious, doesn't it! Column B > Column A.

But if you try $x = \frac{1}{2}$, you get

| $\dfrac{1}{2}$ | $\dfrac{1}{4}$ |

Column A > Column B!

Depending upon the value selected for $x$, the quantity in Column A can be greater or less than the quantity in Column B. The correct choice is (D).

**PROBLEM**

| **Column A** | **Column B** |
|---|---|
| A pound of apples costs 89 cents. | |
| A pound of pears costs 99 cents. | |
| The number of apples in a pound | The number of pears in a pound |

**SOLUTION**

Although a pound of pears costs more than a pound of apples, that does not mean that there are more pears than apples in a pound. There is no way to know how many pears or apples make up a pound. The correct choice is (D).

**PROBLEM**

| **Column A** | **Column B** |
|---|---|
| A rectangle has an area of 8. | |
| The perimeter of the rectangle | 12 |

**SOLUTION**

The centered information tells you the area of the rectangle. Since the area is 8, the length could be 4, and the width would be 2. If these are the measurements, then the perimeter would equal 12. But could the length and width have other values? What if the length is 8 and the width is 1? The area would still be 8, but the perimeter would be 18. We have found one set of values which makes the quantities in Column A and Column B equal (4 and 2) and another set which makes the quantities not equal (8 and 1). The correct choice is (D).

| STEP 5 | Estimation is useful when the numbers in the problem are close to any number that is easy to work with like $1/2$ or 1, or when an approximation will do. |

**PROBLEM**

| Column A | Column B |
| --- | --- |
| $\dfrac{3}{8} + \dfrac{3}{7}$ | 1 |

**SOLUTION**

In this problem, you have to compare the sum of two fractions, with different denominators, to 1. One way to do this is to change both fractions so that they have a common denominator and then do the arithmetic. A faster way is to estimate values. In Column A, $3/8$ is close to $1/2$, but a little less. $3/7$ is also close to $1/2$, but a little less. Since both numbers are each a little less than $1/2$, their sum must be less than 1, the quantity in Column B. The correct choice is (B).

**PROBLEM**

| Column A | Column B |
| --- | --- |
| (16.8) (.51) | (8.4) (.99) |

**SOLUTION**

Each column contains a multiplication problem with positive numbers, but comparing parts doesn't work since

|  | Column A |  | Column B |
| --- | --- | --- | --- |
|  | 16.8 | > | 8.4 |
| but | .51 | < | .99 |

Even if you switch the numbers around, it still won't help:

|  |  |  |
|---|---|---|
| 16.8 | > | .99 |
| but .51 | < | 8.4 |

You may think that you are forced to multiply, but you don't have to if you notice that .51 is a little more than .5 and .99 is a little less than 1. Using these estimates will help you work out the problem without doing all the multiplication in the original problem! (You may find it faster to multiply 16.8 by $\frac{1}{2}$ instead of .5 .)

Using these estimated values in the columns, you get

| **Column A** | **Column B** |
|---|---|
| $(16.8)\left(\dfrac{1}{2}\right)$ | $(8.4)(1)$ |

which becomes

| 8.4 | 8.4 |
|---|---|

*But remember that you are using estimated values!* The real value in Column A is a little more than 8.4 since you really had to multiply 16.8 by .51. The real value in Column B is a little less than 8.4 since you really had to multiply 8.4 by .99. The correct choice is (A).

| STEP 6 | If figures are not drawn to scale or do not look accurate, do not use them to help you solve the problem. These types of figures can throw you off and cause you to select the wrong answer. |
|---|---|

**PROBLEM**

| **Column A** | **Column B** |
|---|---|
| Line $\ell1$ is parallel to line $\ell2$. | |
| Line $\ell2$ is parallel to line $\ell3$. | |
| The distance between $\ell1$ and $\ell2$ | The distance between $\ell2$ and $\ell3$ |

**SOLUTION**

This problem involves parallel lines, but there is no figure. Draw a figure that fits the description in the problem, and then try to draw another figure that also fits the description but shows a different relationship. On the next page are two acceptable figures for this problem. Note that if you draw only one figure, you will get an incorrect answer.

In the first figure, the lines are evenly spaced so the distance between lines $\ell 1$ and $\ell 2$ is the same as the distance between lines $\ell 2$ and $\ell 3$. If this is the only figure you draw, you would choose (C) as your answer.

$\ell 1$
_____

$\ell 2$
_____

$\ell 3$
_____

In this second figure, the distance between lines $\ell 1$ and $\ell 2$ is greater than the distance between lines $\ell 2$ and $\ell 3$. If this is the only figure you draw, you would choose (B) as your answer.

$\ell 1$
_____

$\ell 2$
_____

$\ell 3$
_____

But if you draw both figures, you will choose the correct answer. The correct choice is (D).

The following questions should be completed to further reinforce what you have just learned.

---

Each of the following questions consist of two quantities, one in Column A and one in Column B. You are to compare the two quantities and on the answer sheet blacken space

(A) if the quantity in Column A is greater;

(B) if the quantity in Column B is greater;

(C) if the two quantities are equal;

(D) if the relationship cannot be determined from the information given.

**NOTES:**

1.  In certain questions, information concerning one or both of the quantities to be compared is centered above the two columns.

2.  In a given question, a symbol that appears in both columns represents the same thing in Column A as it does in Column B.

3.  Letters such as $x$, $n$, and $k$ stand for real numbers.

---

## • PROBLEM 2-1

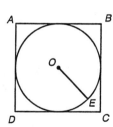

Circle *O* is inscribed in square *ABCD*; $\overline{OE}$ = 4.

Area of square *ABCD*          65

## SOLUTION:

**(B)**   Since the radius $\overline{OE}$ = 4, the diameter of the circle is 8. The diameter equals the side of the square.

The area of the square = $s^2 = 8^2 = 64$.

The correct answer is (B).

## • PROBLEM 2-2

The smallest positive integer,          66
when divided by 4, 5, and 6, that
will always leave a remainder of 3

## SOLUTION:

**(B)**   We need to find the least common multiple for 4, 5, and 6. This number is the lowest number that 4, 5, and 6 can be divided into. This number is 60.

If we want to always have a remainder of 3, we add 3 to the LCM to get 63.

The answer is (B).

• **PROBLEM 2-3**

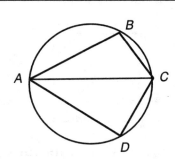

Quadrilateral *ABCD* is inscribed in the circle.

$\angle A + \angle C$                    $\angle B + \angle D$

## SOLUTION:

**(C)**    In a quadrilateral inscribed in a circle, the opposite angles are always supplementary.

$$\angle A + \angle C = \angle B + \angle D = 180°$$

The correct choice is (C).

• **PROBLEM 2-4**

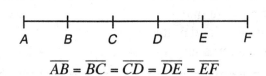

$$\overline{AB} = \overline{BC} = \overline{CD} = \overline{DE} = \overline{EF}$$

$\overline{BD}$ is what % of $\overline{AE}$ ?                    50%

## SOLUTION:

**(C)**    Line segment $\overline{AE}$ is divided into $\overline{AB} + \overline{BC} + \overline{CD} + \overline{DE}$.

Let each segment = $x$

$\overline{AB} = x$

$\overline{BD} = 2x$

$\overline{AE} = 4x$

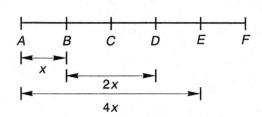

To find out what percentage $\overline{BD}$ is of $\overline{AE}$, we divide $\overline{BD}$ by $\overline{AE}$, multiply by 100, and add a percent sign.

$$\frac{2x}{4x} \times 100 = 50\%$$

The correct answer is (C).

## • PROBLEM 2–5

$$b < 0$$
$$a - b = -b$$

| $a$ | $b$ |

### SOLUTION:

**(A)**   If $b < 0$ and $a - b = -b$, then solving for $a$ we find $a = 0$. $b$ must be negative. Hence, $a > b$.

The correct choice is (A).

## • PROBLEM 2–6

| A boy is numbering the pages of a book by hand. In numbering pages 1 through 100, the number of times the boy will write the number 8 | 19 |

### SOLUTION:

**(A)**   The number of times the boy will write the number 8 is determined as follows:

8, 18, 28, 38, 48, 58, 68, 78, 80, 81, 82, 83, 84, 85, 86, 87, 88, 89, 98

Eighteen of these numbers have one eight, while the number 88 has two eights. Altogether the boy will write 20 eights.

The correct choice is (A).

## • PROBLEM 2-7

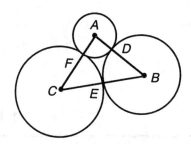

The 3 circles are externally tangent to each other.
Their radii are 4, 5, and 7.

The perimeter of triangle *ABC*                    30

## SOLUTION:

**(A)**    *A* has a radius of 4.

*B* has a radius of 5.

*C* has a radius of 7.

The perimeter of the triangle is found by adding the sides.

$AC = AF + FC = 4 + 7$

$AB = AD + DB = 4 + 5$

$CB = CE + EB = 7 + 5$

The perimeter is therefore equal to

$(4 + 7) + (4 + 5) + (7 + 5) = 11 + 9 + 12 = 32.$

The correct answer is (A).

## • PROBLEM 2-8

$$\frac{\left(\frac{4}{399}\right)\left(\frac{5}{399}\right) + \left(\frac{2}{399}\right)\left(\frac{7}{399}\right) + \left(\frac{6}{399}\right)\left(\frac{1}{399}\right)}{\left(\frac{2}{399}\right)\left(\frac{4}{399}\right)\left(\frac{5}{399}\right)}$$                    399

## SOLUTION:

**(C)**   The key to answering this problem is to simplify.

$$\frac{\left(\dfrac{4}{399}\right)\left(\dfrac{5}{399}\right)+\left(\dfrac{2}{399}\right)\left(\dfrac{7}{399}\right)+\left(\dfrac{6}{399}\right)\left(\dfrac{1}{399}\right)}{\left(\dfrac{2}{399}\right)\left(\dfrac{4}{399}\right)\left(\dfrac{5}{399}\right)}$$

$$\frac{\dfrac{20}{399^2}+\dfrac{14}{399^2}+\dfrac{6}{399^2}}{\dfrac{40}{399^3}}=\frac{\dfrac{40}{399^2}}{\dfrac{40}{399^3}}$$

In order to divide by

$$\frac{40}{399^3},$$

we must multiply by its reciprocal

$$\frac{399^3}{40}.$$

Therefore,

$$\frac{\dfrac{40}{399^2}}{\dfrac{40}{399^3}}=\frac{40}{399^2}\times\frac{399^3}{40}$$

Simplifying we see that this reduces down to 399

The correct answer is (C).

### • PROBLEM 2-9

| 6 is $\dfrac{6}{5}\%$ of $y$. | |
|---|---|
| $y$ | 400 |

## SOLUTION:

**(A)**

$$6 = \frac{6}{5}\% \, y$$

Since $\frac{6}{5} = 1.2$,

then $6 = 1.2\% \, y$.

To change a percent to a decimal we move the decimal point to the left two places and drop the % sign.

$$6 = .012y$$

Solving for $y$ we get

$$y = \frac{6}{.012} = 500.$$

The correct choice is (A).

### • PROBLEM 2-10

$$4 + \frac{4}{.4} + \frac{.4}{4} = Z$$

| $Z$ | 13 |
|-----|-----|

## SOLUTION:

**(A)** In any fraction, you may multiply the numerator and the denominator by any non-zero number without changing its value.

We can multiply the top and bottom of each fraction by 10.

$$4 + \frac{4}{.4} + \frac{.4}{4} = Z$$

$$4 + \frac{40}{4} + \frac{4}{40} = Z$$

$$4 + 10 + \frac{1}{10} = Z$$

$$14\frac{1}{10} = Z$$

The correct answer is (A).

### • PROBLEM 2–11

$$XY = 1, X > 1$$

| | |
|---|---|
| $X$ | $Y$ |

## SOLUTION:

**(A)**  If $XY = 1$ and $X > 1$, then

$Y = \dfrac{1}{x}$ must be less than 1.

$$Y = \frac{1}{x} < 1$$

Therefore, $X > Y$.

The answer is (A).

### • PROBLEM 2–12

$$1 > N > 0$$

| | |
|---|---|
| $N^{10}$ | $N^{12}$ |

## SOLUTION:

**(A)**  $1 > N > 0$ means that $N$ is a positive fraction between 0 and 1. Positive fractions between 0 and 1 become *smaller*, the more times they are multiplied by themselves. Thus, $N^{12}$ is smaller than $N^{10}$ for this problem.

Example:  $\left(\dfrac{1}{2}\right)^{12} < \left(\dfrac{1}{2}\right)^{10}$

The correct answer is (A).

### • PROBLEM 2–13

| | |
|---|---|
| $(2.17)(682)$ | $(196)(6.7)$ |

## SOLUTION:

**(A)**    If we compare (2.17) (682) to (196) (6.7), we see that (2.17) (682) is approximately (2) (680) = 1,360 while (196) (6.7) is approximately (200) (6) = 1,200.

Therefore, (2.17) (682) > (196) (6.7).

The correct choice is (A).

### • PROBLEM 2-14

$$\frac{\dfrac{1}{2}+\dfrac{1}{4}}{\dfrac{1}{3}+\dfrac{1}{5}} \qquad\qquad 1$$

## SOLUTION:

**(A)**

$$\frac{\dfrac{1}{2}+\dfrac{1}{4}}{\dfrac{1}{3}+\dfrac{1}{5}}$$

In order to solve this problem, we must simplify the numerator and the denominator.

$$\text{numerator:} \quad \frac{1}{2}+\frac{1}{4}=\frac{2}{4}+\frac{1}{4}=\frac{3}{4}$$

$$\text{denominator:} \quad \frac{1}{3}+\frac{1}{5}=\frac{5}{15}+\frac{3}{15}=\frac{8}{15}$$

We can now rewrite

$$\frac{\dfrac{1}{2}+\dfrac{1}{4}}{\dfrac{1}{3}+\dfrac{1}{5}} \text{ as } \frac{\dfrac{3}{4}}{\dfrac{8}{15}}.$$

In order to simplify this further, we must remember that dividing by $8/15$ is the same as multiplying by $15/8$. We then have

$$\frac{\dfrac{3}{4}}{\dfrac{8}{15}} = \frac{3}{4} \times \frac{15}{8} = \frac{45}{32}.$$

The correct choice is (A).

## • PROBLEM 2-15

| $(6.9)(8.9)$ | $\sqrt{(50)(82)}$ |

**SOLUTION:**

**(B)** In order to compare $(6.9)(8.9)$ with $\sqrt{(50)(82)}$, we must simplify $\sqrt{(50)(82)}$. Note that

$$\sqrt{(50)(82)} = \sqrt{50} \times \sqrt{82}$$

$$\sqrt{50} \approx 7 \text{ and } \sqrt{82} \approx 9$$

Therefore, $\sqrt{50}\,\sqrt{82} \approx 7 \times 9$.

Since $7 > 6.9$

and $9 > 8.9$

We conclude that $(6.9)(8.9) < \sqrt{(50)(82)}$.

The correct choice is (B).

## • PROBLEM 2-16

| $\dfrac{(5)(0)}{(19)(3)(21)}$ | $(-3)^7 + (9)(3)^5$ |

**SOLUTION:**

**(C)**

$$\frac{(5)(0)}{(19)(3)(21)} = 0$$

because any number multiplied by zero will equal zero.

$$(-3)^7 + 9\,(3)^5 = (-3)^7 + 3^2\,(3)^5$$
$$= (-3)^7 + (3)^7$$
$$= 0$$

The correct choice is (C).

## • PROBLEM 2–17

| | |
|---|---|
| $A + B = 18$ | |
| What is the maximum value of $AB$? | 70 |

## SOLUTION:

(A)    The maximum value of $AB$ occurs when $A = B = 9$. This gives us

$AB = 9 \times 9 = 81$.

Since $81 > 70$, the correct answer is (A).

## • PROBLEM 2–18

| | |
|---|---|
| $(3)^2\,(4)^3$ | $(\pi)^2\,(2)^6$ |

## SOLUTION:

(B)

$$4 = 2^2$$
$$4^3 = (2^2)^3 = 2^6$$

$(3)^2\,(2)^6$ is $< (\pi)^2\,(2)^6$, since 3 is less than 3.14 which is the value of

$\pi$.

The correct choice is (B).

## • PROBLEM 2–19

| | |
|---|---|
| $-\dfrac{7}{8}$ | $-\dfrac{8}{7}$ |

## SOLUTION:

**(A)**

$$\frac{8}{7} > \frac{7}{8}$$

Multiplying both sides by $-1$ we get

$$-\frac{8}{7} < -\frac{7}{8}.$$

The answer is (A).

### • PROBLEM 2-20

$$\frac{x}{y} = 7$$

| $x$ | $y$ |

## SOLUTION:

**(D)**  If

$\frac{x}{y} = 7$, either $x > y$ or $y > x$.

$x > y$ in the case where $x = 7$ , $y = 1$.

$y > x$ in the case where $x = -7$, $y = -1$.

Hence, the answer is (D).

### • PROBLEM 2-21

The sum of three consecutive even
numbers is 42.

| First number | 11 |

## SOLUTION:

**(A)**  Let

$$x = \text{first number}$$

$$x + 2 = \text{second number}$$

$$x + 4 = \text{third number}$$

The sum of these numbers is 42.

$$x + (x + 2) + (x + 4) = 42$$

$$3x + 6 = 42$$

$$3x = 36$$

$$x = 12$$

12 is the first number and is greater than 11.

The correct choice is (A).

## • PROBLEM 2–22

Mary has 2,500 pennies. Ellen has 50 quarters.

| The amount of money Mary has | The amount of money Ellen has |
|---|---|

## SOLUTION:

**(A)**   If Ellen has 50 quarters, then the amount in pennies is

$$(50)\,(25¢) = 1,250¢.$$

Since Mary has 2,500 pennies, she has 2,500¢.

Therefore, the amount of money Mary has (2,500¢) is greater than the amount of money Ellen has (1,250¢).

The correct choice is (A).

## • PROBLEM 2–23

| 35% of 7 | 0.7 of 35 |
|---|---|

## SOLUTION:

**(B)**

$$35\% \text{ of } 7 = .35 \times 7 = 2.45$$

$$0.7 \text{ of } 35 = .7 \times 35 = 24.5$$

Therefore, 35% of 7 < 0.7 of 35.

The correct choice is (B).

### • PROBLEM 2-24

| $(1 - \sqrt{2})(1 - \sqrt{2})$ | $(1 - \sqrt{2})(1 + \sqrt{2})$ |
|---|---|

**SOLUTION:**

**(A)**

$$(1 - \sqrt{2})(1 - \sqrt{2}) = 1 - \sqrt{2} - \sqrt{2} + (\sqrt{2})(\sqrt{2})$$
$$= 1 - 2\sqrt{2} + \sqrt{4}$$
$$= 1 - 2\sqrt{2} + 2$$
$$= 3 - 2\sqrt{2}$$

Similarly:

$$(1 - \sqrt{2})(1 + \sqrt{2}) = 1 - \sqrt{2} + \sqrt{2} - (\sqrt{2})(\sqrt{2})$$
$$= 1 - \sqrt{4}$$
$$= 1 - 2$$
$$= -1$$

Hence, $(1 - \sqrt{2})(1 - \sqrt{2}) > (1 - \sqrt{2})(1 + \sqrt{2})$.

The correct answer is (A).

### • PROBLEM 2-25

| $y = x - 2$ | |
|---|---|
| $y + 3$ | $x - 1$ |

**SOLUTION:**

**(A)**  The value of $y + 3$ is obtained by adding 3 to both sides of the equation $y = x - 2$.

$$y + 3 = x - 2 + 3$$

$$y + 3 = x + 1$$

For any value of $x$, $x + 1 > x - 1$, thus $y + 3 > x - 1$.

The correct choice is (A).

## • PROBLEM 2-26

| The average of 5 consecutive numbers where $x$ is the central number | The average of 3 consecutive numbers where $x$ is the central number |
|---|---|

### SOLUTION:

**(C)**    The average of 5 consecutive numbers where $x$ is the central number appears as follows:

$$\frac{(x - 2) + (x - 1) + (x) + (x + 1) + (x + 2)}{5}$$

Adding like terms, we simplify to get

$$\frac{5x}{5} = x$$

The average of 3 consecutive numbers where $x$ is the central number can be written as follows:

$$\frac{(x - 1) + (x) + (x + 1)}{3}$$

Combining like terms, we get

$$\frac{3x}{3} = x.$$

Thus, the average of 5 consecutive numbers where $x$ is the central number *is equal to* the average of 3 consecutive numbers where $x$ is the central number.

The correct answer is (C).

## • PROBLEM 2–27

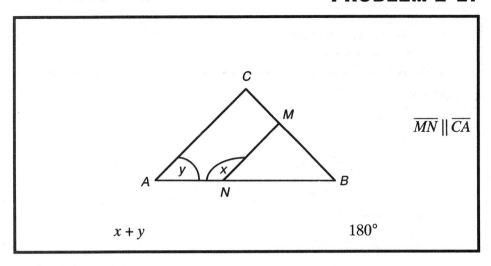

$\overline{MN} \parallel \overline{CA}$

| $x + y$ | $180°$ |

### SOLUTION:

**(C)**   If $\overline{MN} \parallel \overline{CA}$, then

$$\angle MNA + \angle CAB = 180°$$

$$x + y = 180°$$

The correct choice is (C).

## • PROBLEM 2–28

| 30% of 40 | 40% of 30 |

### SOLUTION:

**(C)**

30% of 40 = .30 × 40 = 12.00

40% of 30 = .40 × 30 = 12.00

Therefore, 30% of 40 = 40% of 30.

The correct choice is (C).

## • PROBLEM 2–29

| | |
|---|---|
| The difference between $a + b$ and $a - b$ | $b$ |

### SOLUTION:

**(D)**

$$(a + b) - (a - b) = a + b - a + b$$

$$= 2b$$

If $b > 0$, then $2b > b$.

If $b = 0$, then $2b = b$.

If $b < 0$, then $2b < b$.

Hence, the answer cannot be determined.

The correct choice is (D).

## • PROBLEM 2–30

| | |
|---|---|
| $(2 + (6 \times 3) - 4)$ | $((2 + 6) \times (3 - 4))$ |

### SOLUTION:

**(A)**    This problem is a good example of using the proper order of operations. Operations in parentheses and exponents are performed before multiplication and division which are performed before addition and subtraction.

$$(2 + (6 \times 3) - 4) = (2 + (18) - 4)$$

$$= 20 - 4$$

$$= 16$$

$$((2 + 6) \times (3 - 4)) = 8 \times -1$$

$$= -8$$

Hence, $16 > -8$.

The correct choice is (A).

## • PROBLEM 2–31

Let $x > 0$.

| | |
|---|---|
| $x$ | $x^2$ |

## SOLUTION:

**(D)**   Let $x > 0$

If $x > 1$, then $x < x^2$.

On the other hand if $0 < x < 1$, $x^2 < x$.

No determination can be made.

The correct choice is (D).

## • PROBLEM 2–32

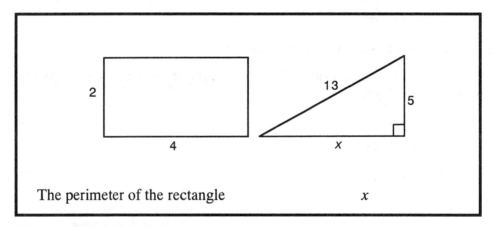

The perimeter of the rectangle                                    $x$

## SOLUTION:

**(C)**   The perimeter of the rect-angle is $L + L + W + W$. Since $L = 4$ and $W = 2$, the perimeter equals

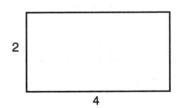

$$4 + 4 + 2 + 2 = 12.$$

We can apply the Pythagorean Theorem to determine the value of $x$.

$$x^2 + y^2 = z^2$$
$$x^2 + 5^2 = 13^2$$

$$x^2 + 25 = 169$$

$$x^2 = 169 - 25$$

$$x^2 = 144$$

$$x = 12$$

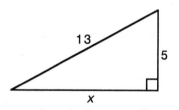

Both columns are equal.

The correct choice is (C).

## • PROBLEM 2–33

| $y$ is negative. | |
| --- | --- |
| $y + 1$ | $-2y + 1$ |

## SOLUTION:

**(B)**    $y$ is negative.

Since $y < 0$, $-2y > 0$ and so $-2y > y$.

If we add 1 to both sides of the inequality, the sense of the inequality is not changed so that

$$-2y + 1 > y + 1.$$

Therefore, column (B) is greater.

The answer is (B).

## • PROBLEM 2–34

| The sum of three consecutive numbers is $3x$. | |
| --- | --- |
| The smallest of the three | $x$ |

## SOLUTION:

**(B)**    The sum of three consecutive numbers is $3x$.

Let         $y$ = first number

$y + 1$ = second number

$y + 2 =$ third number

$y + (y + 1) + (y + 2) = 3x$

$$3y + 3 = 3x$$

$$y + 1 = x$$

Therefore, the smallest number $y < y + 1$, so $y < x$.

The correct choice is (B).

### • PROBLEM 2-35

One boy sleeps 14 hours per day (1 month = 30 days).

| Number of hours that the boy sleeps in 3 weeks | Number of hours that the boy doesn't sleep in a month |
|---|---|

## SOLUTION:

**(B)**  The number of hours that the boy sleeps in three weeks is given by

14 hours/day × 7 days/week × 3 weeks = 294 hrs.

The number of hours the boy does not sleep in a month is

10 hours/day × 30 days/month × 1 month = 300 hours

The correct answer is therefore (B).

### • PROBLEM 2-36

Let $ab = 5$ and $(a + b)^2 = 12$.

| $a^2 + b^2$ | 2 |
|---|---|

## SOLUTION:

**(C)**  Let

$$ab = 5$$

$$(a + b)^2 = 12$$

$$(a + b)^2 = a^2 + 2ab + b^2$$

$$(a + b)^2 - 2ab = a^2 + b^2$$
$$12 - 2(5) = a^2 + b^2$$
$$2 = a^2 + b^2$$

The numbers in both columns are the same.

The correct answer is (C).

## • PROBLEM 2-37

$$0 < X < 2$$

| $X^2$ | $X^3$ |

**SOLUTION:**

**(D)**   $0 < X < 2$ means that $x$ is a number between 0 and 2.

Break up the given interval into several smaller intervals.

In the interval $0 < x < 1$ (positive fractions):

$X^2 > X^3.$

When $X = 1$:

$X^2 = X^3.$

In the interval $1 < X < 2$ ($X$-values between 1 and 2):

$X^2 < X^3.$

The answer cannot be determined since we have shown that $X^2$ may be greater than, equal to, or less than $X^3.$

The answer is (D).

## • PROBLEM 2-38

The three-digit number 4 ■ 4 is divisible by 8.

| ■ | 8 |

## SOLUTION:

**(B)**    The three-digit number 4 ■ 4 is divisible by 8.

Since 400 is divisible by 8, it is sufficient to examine those two-digit numbers ending in 4 that are divisible by 8. They are 24 and 64. Thus ■ = 2 or 6. In either case, that is less than 8. So Column B is bigger.

The correct answer is (B).

## • PROBLEM 2-39

| The values of $y$ in $y^2 + 12y + 27 = 0$ | The values of $x$ in $x^2 - 12x + 27 = 0$ |
|---|---|

## SOLUTION:

**(B)**

$$y^2 + 12y + 27 = 0 \qquad x^2 - 12x + 27 = 0$$
$$(y + 9)(y + 3) = 0 \qquad (x - 9)(x - 3) = 0$$
$$y + 9 = 0 \qquad x - 9 = 0$$
$$y = -9 \qquad x = 9$$
$$y + 3 = 0 \qquad x - 3 = 0$$
$$y = -3 \qquad x = 3$$

The value of $y$ in Column A is negative for both solutions. The value of $x$ in Column B is positive. Therefore, $y$ is less than $x$.

The correct choice is (B).

• **PROBLEM 2–40**

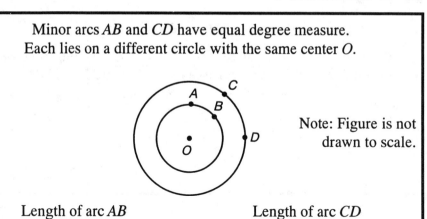

Minor arcs *AB* and *CD* have equal degree measure.
Each lies on a different circle with the same center *O*.

Note: Figure is not drawn to scale.

Length of arc *AB*          Length of arc *CD*

**SOLUTION:**

**(B)**    Minor arcs *AB* and *CD* have equal degree measure.

By definition the arc length equals the angular measure times the radius. Since *CD* and *AB* have equal angles, but *CD* lies on the circle with the larger radius, *CD* has the larger arc length.

The correct choice is (B).

• **PROBLEM 2–41**

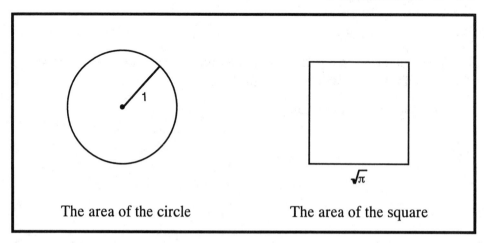

The area of the circle          The area of the square

**SOLUTION:**

**(C)**    The area of a circle is $\pi r^2$.

The area of a circle having a radius of 1 is $\pi (1)^2 = \pi$.

The area of a square is $s^2$, where $s$ = length of a side.

The area of a square having a side equal to $\sqrt{\pi}$ is

$$(\sqrt{\pi})^2 = \sqrt{\pi} \times \sqrt{\pi} = \pi.$$

Hence, the area of the circle is equal to the area of the square.

The correct choice is (C).

## • PROBLEM 2-42

| The average of 18, 20, 22, 24, 26 | The average of 17, 19, 21, 23, 25, 27 |
|---|---|

## SOLUTION:

**(C)**    The average of the numbers in Column A is

$$\frac{18 + 20 + 22 + 24 + 26}{5} = \frac{110}{5} = 22.$$

Likewise, the average of the numbers in Column B is

$$\frac{17 + 19 + 21 + 23 + 25 + 27}{6} = \frac{132}{6} = 22.$$

So, the quantities are equal in both columns.

The correct choice is (C).

## • PROBLEM 2-43

| List A | List B |
|---|---|
| 10 | 20 |
| 20 | 25 |
| 30 | 30 |
| 40 | 35 |
| 50 | 40 |
| The average of the numbers in List A | The average of the numbers in List B |

## SOLUTION:

**(C)** The average of the numbers in List A is

$$\frac{10 + 20 + 30 + 40 + 50}{5} = \frac{150}{5} = 30.$$

The average of the numbers in List B is

$$\frac{20 + 25 + 30 + 35 + 40}{5} = \frac{150}{5} = 30.$$

Hence, the quantities are equal in both columns.

The correct choice is (C).

### • PROBLEM 2–44

| 0 | The largest even integer smaller than 2 |
|---|---|

## SOLUTION:

**(C)** An even number is a number that is divisible by 2. The number 0 is an integer, and it is also an even integer because it is divisible by 2. Zero is also the largest even integer smaller than 2.

Thus, the quantities in both columns are equal.

The correct choice is (C).

### • PROBLEM 2–45

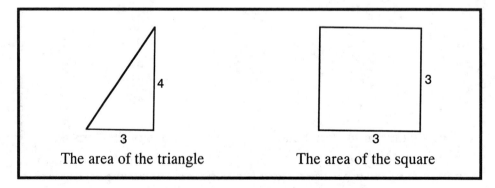

| The area of the triangle | The area of the square |
|---|---|

## SOLUTION:

**(B)**    The area of the triangle is

$$\frac{1}{2} (b) (h).$$

We are given that $b = 3$, and $h = 4$. The area is therefore equal to

$$\frac{1}{2} (b) (h) = \frac{1}{2} (3) (4) = 6.$$

The area of a square is (side)$^2$. Since $s = 3$, the area is $(3)^2 = 9$.

Therefore, the area of the triangle is less than the area of the square.

The correct choice is (B).

## • PROBLEM 2–46

Of the 50 students in a social studies class,
exactly 60% are passing.

| The number who are failing | 40 |
|---|---|

## SOLUTION:

**(B)**    In a class of 50 students, if 60% are passing then 100% − 60% = 40% are failing.

$$40\% \text{ of } 50 = .40 \times 50$$

$$= 20$$

We changed the percent to a decimal and multiplied.

The number failing is 20, which is less than 40.

The correct choice is (B).

## • PROBLEM 2-47

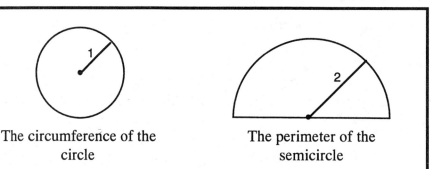

The circumference of the circle

The perimeter of the semicircle

### SOLUTION:

**(B)** The circumference of a circle is $2\pi r$. If the radius is 1, then the circumference is $2\pi (1) = 2\pi$.

The perimeter of a semicircle can be found in the following way. The curved portion is just half the circumference of the full circle. This is

$$\frac{1}{2} (2\pi r) = \frac{1}{2} (2\pi \times 2) = 2\pi.$$

To this must be added the length of the diameter which is 4.

The perimeter of the semicircle is $2\pi + 4$.

Since $2\pi + 4 > 2\pi$, the perimeter of the semicircle is greater than the circumference of the circle.

The correct choice is (B).

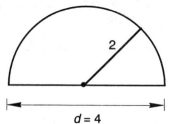

$d = 4$

## • PROBLEM 2-48

| $x + x + x$ | $x^3$ |
|---|---|

## SOLUTION:

**(D)**

$$x + x + x = 3x$$

If $\quad x = 0,$

then $\quad 3x = 3(0) = 0$

and $\quad x^3 = 0.$

In this case, $3x = x^3$.

But if $\quad x = 1,$

then $\quad 3x = 3(1) = 3$

and $\quad x^3 = 1.$

In this case, $3x \neq x^3$.

We cannot determine from the given information which quantity is greater.

The answer is (D).

### • PROBLEM 2–49

| $6x - 3$ | $2x - 1$ |
|---|---|

## SOLUTION:

**(D)** $\quad 6x - 3$ can be factored to give $3 (2x - 1)$. This is three times the value given in Column 2.

If $x > 0$, then Column A is greater than Column B.

However, if $x \leq 0$, then Column A is less than Column B.

We cannot determine from the information given which quantity is greater.

The answer is (D).

### • PROBLEM 2-50

| The sum of all angles of a polygon whose sides are all equal | The sum of all angles of a square |
|---|---|

**SOLUTION:**

**(D)** The sum of all angles of a square is

$$90° + 90° + 90° + 90° = 360°.$$

The sum of all angles of any polygon with equal sides will vary. For example, an equilateral triangle has the sum of all angles equal to $180°$, while a hexagon with equal sides has the sum of the angles equal to $720°$. Thus, it is not possible to compare the results of the two columns.

The answer is (D).

### • PROBLEM 2-51

| The product of the roots of the equation $x^2 + 3x + 2 = 0$ | −1 |
|---|---|

**SOLUTION:**

**(A)** To find the roots of

$$x^2 + 3x + 2 = 0,$$

we must factor:

$$x^2 + 3x + 2 = 0$$
$$(x + 2)(x + 1) = 0$$
$$(x + 2) = 0 \qquad (x + 1) = 0$$
$$x = -2 \qquad\qquad x = -1$$

The roots −2 and −1 have a product of $(-2)(-1) = 2$. Hence, the product of the roots of the equation

$$x^2 + 3x + 2 = 0$$

is greater than −1.

The answer is (A).

• **PROBLEM 2–52**

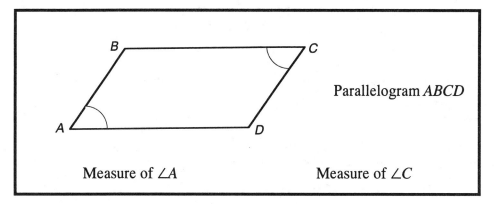

Measure of $\angle A$          Measure of $\angle C$

## SOLUTION:

**(C)**    The diagram shows that $\angle A$ and $\angle C$ are opposite angles of parallelogram *ABCD*. Opposite angles in a parallelogram are equal. The measure of $\angle A$ = measure of $\angle C$.

The correct choice is (C).

• **PROBLEM 2–53**

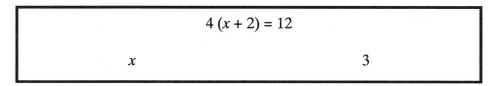

## SOLUTION:

**(B)**

$$4(x + 2) = 12$$

Using the distribution law, we can solve for *x*.

$$4x + 8 = 12$$

$$4x = 4$$

$$x = 1$$

So, $x = 1$ and $3 > 1$. Column B is greater than Column A.

The answer is (B).

### • PROBLEM 2-54

$$x = 5, y = -3$$

$(x + y)^2$ $\qquad\qquad\qquad\qquad$ $(x - y)^2$

## SOLUTION:

**(B)** To compare $(x + y)^2$ with $(x - y)^2$ when $x = 5$ and $y = -3$, we must solve each expression by substituting $x = 5$ and $y = -3$.

$$(x + y)^2 = (5 + (-3))^2 = (2)^2 = 4$$
$$(x - y)^2 = (5 - (-3))^2 = (8)^2 = 64$$

Column B is therefore larger than Column A.

The correct choice is (B).

### • PROBLEM 2-55

$4.445 \times 10^5$ $\qquad\qquad\qquad\qquad$ $445,000$

## SOLUTION:

**(B)**

$$4.445 \times 10^5 = 4.445 \times 100,000 = 444,500$$

A quick way to find the value of $4.445 \times 10^5$ is to take 4.445 and move the decimal point 5 places to the right (add extra zeros after the five in 4.445 as placeholders).

Since $4.445 \times 10^5 = 444,500 < 445,000$, Column B is greater than Column A.

The correct choice is (B).

### • PROBLEM 2-56

$\sqrt{9}$ $\qquad\qquad\qquad\qquad$ The smaller root of
$x^2 - 2x - 15 = 0$

## SOLUTION:

**(A)**    When we factor the expression

$$x^2 - 2x - 15 = 0,$$

we get

$$(x + 3)(x - 5) = 0.$$

Setting each of these factors to zero, we get

$$(x + 3) = 0 \qquad (x - 5) = 0$$

$$x = -3 \qquad\qquad x = 5$$

The smaller root is $x = -3$.

$$\sqrt{9} = 3$$

Since $-3 < 3$, Column A is larger than Column B.

Hence, the correct answer is (A).

### • PROBLEM 2–57

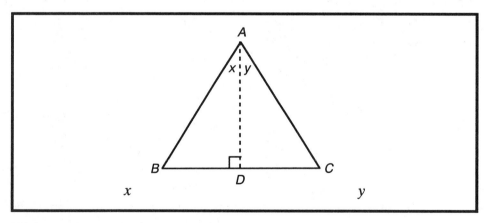

## SOLUTION:

**(D)**    If line segments $\overline{AD}$ and $\overline{BC}$ are perpendicular, then $\angle ADB = \angle ADC = 90°$. Since no other information is given about $\triangle ABD$ or $\triangle ACD$, we cannot infer any relationship between $\angle x$ and $\angle y$.

Hence, the correct choice is (D).

## • PROBLEM 2–58

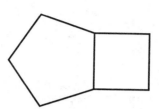

The regular pentagon and the square share a common side.
The perimeter of the pentagon is $12^1/_2$.

Area of the square                                                6

## SOLUTION:

**(A)**    Since all sides of a regular pentagon are equal, we can divide the perimeter of the pentagon ($12^1/_2$) by the number of sides in a pentagon (5).

$$12\frac{1}{2} \div 5 = \frac{25}{2} \div 5$$

$$= \frac{25}{2} \times \frac{1}{5}$$

$$= \frac{25}{10}$$

$$= 2\frac{1}{2}$$

This gives us the length of each side in the pentagon. Since the pentagon and square share a common side, the length of the side of the square is also $2^1/_2$.

The area of the square is

$$(\text{side})^2 = \left(2\frac{1}{2}\right)^2 = \left(\frac{5}{2}\right)^2 = \frac{25}{4} = 6\frac{1}{4}.$$

Hence, the area of the square is $6^1/_4$ which is greater than 6.

The correct choice is (A).

### • PROBLEM 2–59

| | |
|---|---|
| 2 inches is what fraction of a yard? | 2 minutes is what fraction of an hour? |

**SOLUTION:**

**(A)** Since 1 yard = 36 inches, then 2 inches is

$$\frac{2}{36} = \frac{1}{18} \text{ of a yard.}$$

Since 1 hour = 60 minutes, then 2 minutes is

$$\frac{2}{60} = \frac{1}{30} \text{ of an hour.}$$

$$\frac{1}{30} < \frac{1}{18}$$

So Column A is greater than Column B.

The correct choice is (A).

### • PROBLEM 2–60

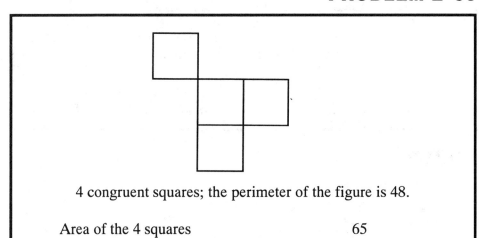

4 congruent squares; the perimeter of the figure is 48.

Area of the 4 squares                          65

**SOLUTION:**

**(B)** Let $s$ = the side of the square. We are given that the perimeter is 48. Since there are 12 sides, each side equals

$$\frac{48}{12} = 4.$$

Hence, the length of one side is 4. The area of one of the squares is

$$(4)^2 = 4 \times 4 = 16 \text{ sq units.}$$

Since there are four squares, the total area is

$$4 \times 16 \text{ sq units} = 64 \text{ sq units.}$$

The area of the 4 squares is 64 sq units which is less than 65.

The correct choice is (B).

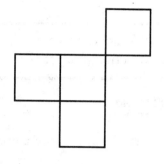

## • PROBLEM 2-61

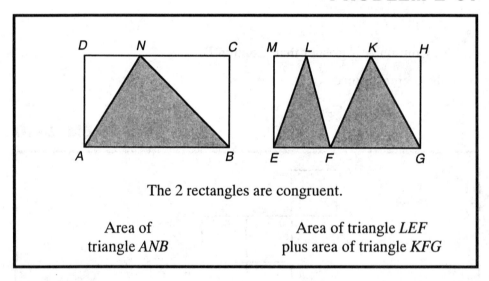

The 2 rectangles are congruent.

| Area of triangle *ANB* | Area of triangle *LEF* plus area of triangle *KFG* |

## SOLUTION:

(C)

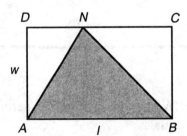

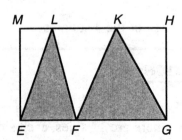

Let    $l$ = length $AB$

   $w$ = width $AD$

The area of triangle $ANB$ equals

$$\frac{1}{2}\ (\text{base})\ (\text{height}) = \frac{1}{2}\ lw.$$

Now we will consider rectangle $EGHM$. Since rectangles $ABCD$ and $EGHM$ are congruent, then the length $EG$ = the length of $AB$ and the length $AD$ = the length of $EM$.

   $EG = AB = l$   $AD = EM = w$

If we set the length $EF = x$, then the length $FG = l - x$.

The area of triangle $LEF$ equals

$$\frac{1}{2}\ (\text{base})\ (\text{height}) = \frac{1}{2}\ xw.$$

Similarly, the area of triangle $KFG$ equals

$$\frac{1}{2}\ (\text{base})\ (\text{height}) = \frac{1}{2}\ (l - x)\ (w)$$

$$= \frac{1}{2}\ lw - \frac{1}{2}\ xw$$

The area of triangle $LEF$ plus triangle $KFG$ equals

$$\frac{1}{2}\ xw + \frac{1}{2}\ lw - \frac{1}{2}\ xw = \frac{1}{2}\ lw.$$

This value is equal to the area of triangle $ANB$.

The correct choice is (C).

## • PROBLEM 2-62

$$a > 0, b > 0, c > 0$$
$$ab = 2$$
$$bc = 9$$
$$ac = 8$$

| $abc$ | 14 |
|---|---|

## SOLUTION:

**(B)**   We are given that

$$ab = 2 \tag{1}$$

$$bc = 9 \tag{2}$$

$$ac = 8 \tag{3}$$

We must first solve equation (1) for $a$.

$$ab = 2$$

$$a = \frac{2}{b}$$

We can then substitute this value for $a$ in (3).

$$ac = 8$$

$$\frac{2}{b} \times c = 8$$

$$2c = 8b$$

$$c = 4b$$

We then substitute $c = 4b$ in equation (2).

$$bc = 9$$

$$b(4b) = 9$$

$$4b^2 = 9$$

$$b^2 = \frac{9}{4}$$

$$b = \frac{3}{2}$$

We can now substitute this numerical value for $b$ in (1).

$$ab = 2$$

$$a\left(\frac{3}{2}\right) = 2$$

$$3a = 4$$

$$a = \frac{4}{3}$$

We are now able to substitute this numerical value for $a$ in (3).

$$ac = 8$$

$$\frac{4}{3}c = 8$$

$$4c = 24$$

$$c = 6$$

$$a \times b \times c = \left(\frac{4}{3}\right)\left(\frac{3}{2}\right)(6) = 12$$

*abc* equals 12 which is less than 14.

The correct answer is (B).

## • PROBLEM 2-63

If $a \times b = 2ab - b$ and $b \times a = 2a - 1$
$$a, b > 1$$

| $\dfrac{a \times b}{b \times a}$ | $\dfrac{b \times a}{a \times b}$ |
|---|---|

## SOLUTION:

**(A)**

$$a \times b = 2ab - b$$

$$b \times a = 2a - 1 \qquad a, b > 1$$

Evaluate Column A

$$\frac{a \times b}{b \times a} = \frac{2ab - b}{2a - 1}$$

$$\frac{a \times b}{b \times a} = \frac{b(2a - 1)}{2a - 1} = b$$

Evaluate Column B

$$\frac{b \times a}{a \times b} = \frac{(2a - 1)}{2ab - b}$$

$$\frac{b \times a}{a \times b} = \frac{(2a - 1)}{b(2a - 1)} = \frac{1}{b}$$

Since we are given that $b > 1$, then $b > \dfrac{1}{b}$.

Hence,

$$\dfrac{a \times b}{b \times a} > \dfrac{b \times a}{a \times b}.$$

The correct choice is (A).

## • PROBLEM 2–64

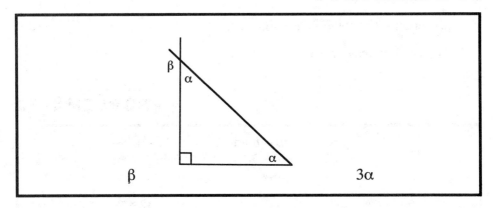

## SOLUTION:

**(C)**    The sum of the angles in a triangle must add up to 180°. In the diagram we have a triangle with one right angle (90°) and two angles of value $\alpha$.

$$\alpha + \alpha + 90° = 180°$$

$$2\alpha + 90° = 180°$$

$$2\alpha = 90°$$

$$\alpha = 45°$$

Since $\alpha$ and $\beta$ are on a straight line, they are supplementary. This means that their sum must equal 180°.

$$\alpha + \beta = 180°$$

Substituting $\alpha = 45$, we find

$$45 + \beta = 180°$$

$$\beta = 135°$$

$3\alpha = 3(45) = 135°$

Hence, $3\alpha = \beta$.

The correct choice is (C).

## • PROBLEM 2–65

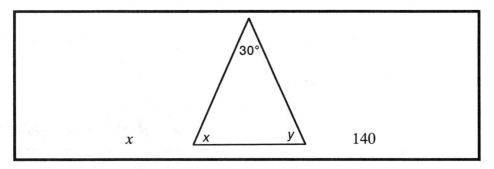

$$x \qquad\qquad\qquad 140$$

### SOLUTION:

**(D)**    The sum of the angles in a triangle must add up to $180°$.

$$x + y + 30° = 180°$$

$$x + y = 180° - 30° = 150°$$

$x$ cannot equal $150°$ but could be greater than $140°$. It could also be less than $140°$. Since the problem gives no additional information about $y$, not enough information is given to determine if $x$ is greater than or less than $140°$.

The correct choice is (D).

## • PROBLEM 2–66

| $n > 2$ | |
|---|---|
| Half of $n^2$ | $n/2$ |

### SOLUTION:

**(A)**    Column A:

$$\text{Half of } n^2 = \frac{1}{2}n^2 = \frac{n^2}{2}$$

We need to compare $\dfrac{n^2}{2}$ from Column A with $\dfrac{n}{2}$ from Column B.

We will do this by comparing $n^2$ with $n$. If $n > 1$, then $n^2 > n$. We know that $n$ must be greater than 1 for this problem, because $n > 2$ was given as true.

Therefore,

$$\frac{n^2}{2} > \frac{n}{2} \text{ for this problem.}$$

The correct choice is (A).

## • PROBLEM 2–67

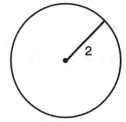

The ratio of the area of the circle to the circumference of the circle

The ratio of the area of the square to the perimeter of the square

## SOLUTION:

**(C)**

For the circle:

$$\frac{\text{area of circle}}{\text{circumference}} = \frac{\pi r^2}{2\pi r} = \frac{\pi 2^2}{2\pi 2} = 1$$

For the square:

$$\frac{\text{area of square}}{\text{perimeter of square}} = \frac{s^2}{4s} = \frac{4^2}{4 \times 4} = 1$$

The two ratios are the same.

The correct choice is (C).

## • PROBLEM 2-68

| Let $n$ be a three-digit number. | |
| --- | --- |
| The sum of the digits of $n$ | 28 |

**SOLUTION:**

**(B)**  The largest any single digit can be is 9. The largest three-digit number is 999. If we add the sum of these three digits, we get

$9 + 9 + 9 = 27$.

Hence, no matter what three-digit number is considered, it will always be less than 28.

The correct choice is (B).

## • PROBLEM 2-69

| $A$ is bigger than $P$, and $B$ is not less than $P$. | |
| --- | --- |
| $A$ | $B$ |

**SOLUTION:**

**(D)**  If $A$ is bigger than $P$, then $A > P$.

If $B$ is not less than $P$, then $B$ is greater than or equal to $P$, or $B \geq P$.

$A > P$

$B \geq P$

Therefore, $A > B$, $A = B$, or $A < B$.

The correct choice is (D).

### • PROBLEM 2-70

> A school baseball team has won
> 6 out of 8 games.
>
> Their record, in percent, if                    75%
> they win 2 more games in a row

### SOLUTION:

**(A)**    If the team wins 2 more games in a row, then the team will have won $6 + 2 = 8$ games out of a total of $8 + 2 = 10$ games.

The ratio 8 games out of 10 is

$$\frac{8}{10} = .80.$$

We can convert this to a percent by moving the decimal point two places to the right.

$.80 = 80\%$

Hence, $80\% > 75\%$.

The correct choice is (A).

### • PROBLEM 2-71

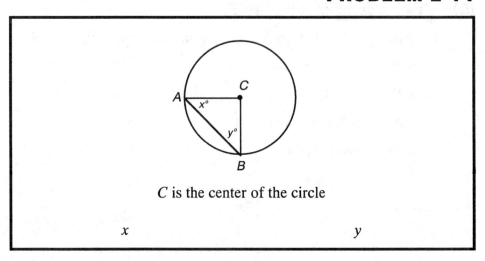

$C$ is the center of the circle

x                                                    y

## SOLUTION:

**(C)** Since $\overline{AC}$ and $\overline{BC}$ are radii of the circle, then triangle $ABC$ is an isosceles triangle. $\overline{AC}$ and $\overline{BC}$ are congruent. In an isosceles triangle the base angles opposite the congruent sides are congruent. Thus, $\angle A$ and $\angle B$ are congruent making $x$ and $y$ equal.

The answer is (C).

## • PROBLEM 2-72

| $\angle A = 50°$ | |
|---|---|
| $\angle A$ | The supplement of the complement of $\angle A$ |

## SOLUTION:

**(B)** Complementary angles add up to 90°.

Supplementary angles add up to 180°.

Using the above information, we can determine the complement of $\angle A$.

$90° - \angle A = 90° - 50° = 40°$ (complement)

We can determine the supplement of the complement as follows:

$180° - 40° = 140°$

Hence, $\angle A = 50°$ which is less than 140°.

The correct choice is (B).

## • PROBLEM 2-73

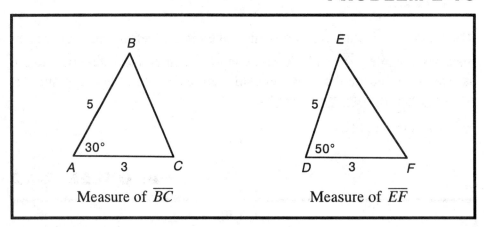

Measure of $\overline{BC}$          Measure of $\overline{EF}$

## SOLUTION:

**(B)**    Imagine $\angle BAC$ and $\angle EDF$ as metal hinges of equal size.

$\overline{BA} = \overline{ED} = 5$ and $\overline{AC} = \overline{DF} = 3$.

The hinge that is spread to the wider angle has the wider opening. Since $\angle EDF$ is opened to 50° and $\angle BAC$ is opened to 30°, then $\overline{EF}$ will be longer than $\overline{BC}$.

The correct choice is (B).

## • PROBLEM 2-74

Automobile $A$ has a 15-gallon tank and gets 20 miles per gallon. Automobile $B$ has a 20-gallon tank, but gets 15 miles per gallon. Both cars go on a 600-mile trip.

The number of tanks          The number of tanks
used by $A$                   used by $B$

## SOLUTION:

**(C)**    Automobile $A$ can go

20 miles/gal × 15 gal on one tank.

20 mi/gal × 15 gal = 300 miles

Automobile $A$ will therefore need

$$\frac{600}{300} = 2 \text{ tanks of gas.}$$

Automobile *B* can go

15 miles/gal × 20 gal on one tank

15 mi/gal × 20 gal = 300 miles

Automobile *B* will need

$$\frac{600}{300} = 2 \text{ tanks of gas.}$$

The number of tanks used by *A* is equal to the number of tanks used by *B*.

The correct choice is (C).

• **PROBLEM 2-75**

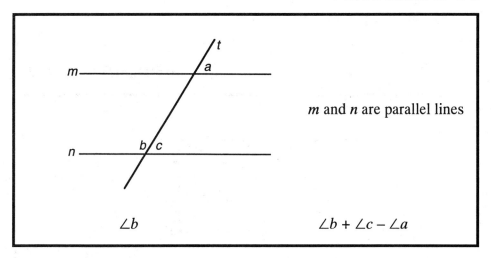

*m* and *n* are parallel lines

∠*b*                    ∠*b* + ∠*c* − ∠*a*

## SOLUTION:

**(C)**   Since *m* and *n* are parallel lines cut by the transversal *t*, then ∠*a* and ∠*c* are corresponding angles. Angle *a* and angle *c* are therefore equal.

We may simplify the expression in Column B by replacing ∠*a* with ∠*c*.

$$\angle b + \angle c - \angle a = \angle b + \angle c - (\angle c)$$

$$= \angle b + 0$$

$$= \angle b$$

Therefore, $\angle b$ is equal to the term in Column B.

The correct choice is (C).

## • PROBLEM 2–76

| $\sqrt{9} + \sqrt{7}$ | $\sqrt{16}$ |
|---|---|

## SOLUTION:

**(A)**

Since $\sqrt{9} = 3$ and $\sqrt{7} > 2$, then it is clear that

$\sqrt{9} + \sqrt{7} > 5$.

But $\sqrt{16} = 4$. So, the quantity in Column A is greater than the quantity in Column B.

The correct choice is (A).

## • PROBLEM 2–77

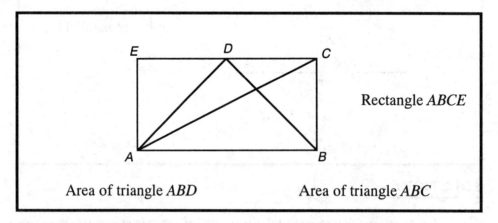

Rectangle *ABCE*

| Area of triangle *ABD* | Area of triangle *ABC* |
|---|---|

## SOLUTION:

**(C)**  The area of a triangle is equal to $\frac{1}{2}$ the base times the height. We can see that both triangles *ABD* and *ABC* have the same base, *AB*. Both triangles also have the same height, *BC*.

Hence, the two areas are equal.

The answer is (C).

### • PROBLEM 2-78

$$3x = 5y; \quad \frac{x}{y} = \frac{a}{b}$$

| $a$ | $b$ |
|---|---|

**SOLUTION:**

**(D)**

$$\frac{x}{y} = \frac{a}{b}$$

Cross multiplying we get

$bx = ay$.

Comparing this to $3x = 5y$, we can see that $b = 3$ and $a = 5$. In this case, $a > b$. But $b = -3$ and $a = -5$ also satisfy the proportion. In this case $a < b$.

Therefore, the correct choice is (D).

### • PROBLEM 2-79

$$0.5x - 0.5y = 3$$

| $x$ | $y$ |
|---|---|

**SOLUTION:**

**(A)** We must solve the given equation for $x$.

$$0.5x - 0.5y = 3$$

$$\frac{1}{2}x - \frac{1}{2}y = 3$$

$$\frac{1}{2}(x - y) = 3$$

$$x - y = 6$$

$$x = y + 6$$

Hence, $x > y$.

The answer is (A).

## • PROBLEM 2-80

| | |
|---|---|
| $wz - xy$ | $0$ |

### SOLUTION:

**(D)**    Since the expression $wz - xy$ does not show any relationship between the variables $w$, $z$, $x$, and $y$, we cannot infer any relationship between the terms in Column A and B.

The correct choice is (D).

## • PROBLEM 2-81

| $m > n > 0$ | |
|---|---|
| $x^m$ | $x^n$ |

### SOLUTION:

**(D)**

$$m > n > 0$$

If $x = 0$, then $x^m = x^n$

If $x > 0$, then $x^m = x^n$

   or     $x^m = x^n$

   or     $x^m < x^n$

If $x < 0$, then $x^m = x^n$

   or     $x^m > x^n$

   or     $x^m < x^n$

The relationship between $x^m$ and $x^n$ cannot be determined.

The correct choice is (D).

• **PROBLEM 2-82**

| | |
|---|---|
| $a(a - c) + b(a - c)$ | $(a + b)(a - c)$ |

**SOLUTION:**

**(C)** $\quad a(a - c) + b(a - c)$

can be simplified by taking out a common factor of $(a - c)$. Thus,

$\quad a(a - c) + b(a - c) = (a + b)(a - c)$.

This is identical to the expression in Column B.

The correct choice is (C).

• **PROBLEM 2-83**

| | |
|---|---|
| $y^5 \times y^4$ | $y^1 \times y^3 \times y^4$ |

**SOLUTION:**

**(A)** We evalute the given autntities by adding the exponents.

$\quad y^5 \times y^4 = y^{(5 + 4)} = y^9 \qquad$ (Column A)

$\quad y^1 \times y^3 \times y^4 = y^{(1 + 3 + 4)} = y^8 \quad$ (Column B)

Since $y^9 > y^8$ when $y > 1$, Column A is greater than Column B.

The correct answer is (A).

• **PROBLEM 2-84**

| | |
|---|---|
| Given a cube with length of a side equal to $d$ units | |
| Surface area of cube | Volume of cube |

**SOLUTION:**

**(D)** The formula for the surface area of a cube is the sum of the area of the 6 faces of the cube. The area of each face is $d(d) = d^2$. The surface area is thus equal to $6d^2$.

The volume of a cube is given by

$$(d) (d) (d) = d^3.$$

If $d > 6$, then $d^3 > 6d^2$

If $d = 6$, then $d^3 = 6d^2$

If $d < 6$, then $d^3 < 6d^2$.

Therefore, no comparison can be made with the given information.

The correct choice is (D).

### • PROBLEM 2-85

---

Neither $p$ nor $q$ are zero.

$$\frac{p}{q} \qquad\qquad\qquad \frac{q}{p}$$

---

## SOLUTION:

**(D)**   If $p > q > 0$, then

$$\frac{p}{q} > \frac{q}{p}.$$

If $p = -1$ and $q = 1$, then

$$\frac{p}{q} = \frac{q}{p}.$$

Hence, not enough information is given to determine the relationship.

The answer is (D).

## • PROBLEM 2-86

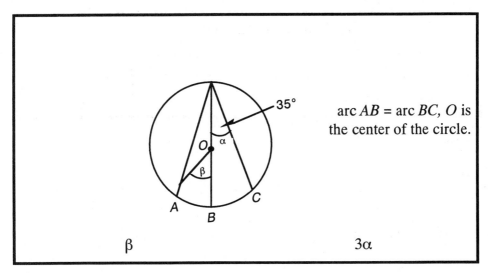

arc $AB$ = arc $BC$, $O$ is the center of the circle.

$\beta$                            $3\alpha$

## SOLUTION:

**(B)**    $\alpha$ is an *inscribed* angle (vertex is on the perimeter of the circle).

$\beta$ is a *central* angle (vertex is at the center of the circle).

Since arc $AB$ = arc $BC$, $\alpha$ and $\beta$ intercept arcs of the same length.

Then $\alpha = \frac{1}{2}\beta$, because the measure of an inscribed angle equals half the measure of a central angle when both angles intercept arcs of equal length.

Solve for $\beta$

$$\alpha = \frac{1}{2}\beta$$

$$\beta = 2\alpha$$

Since $\beta = 2\alpha$ is in Column A, and $3\alpha$ is in Column B, Column B is larger.

The correct answer is (B).

$$x > y > z, z > 0$$

$$\frac{1}{xy} \qquad\qquad \frac{1}{yz}$$

## SOLUTION:

**(B)**

$$x > y > z, z > 0$$

We can see that comparing

$$\frac{1}{xy} \text{ with } \frac{1}{yz}$$

is the same as comparing

$$\frac{1}{x} \text{ with } \frac{1}{z}.$$

This is because $y$ appears in both denominators. (We can only ingnore $y$ in this way, because $y$ is given to be a *positive* number. If $y$ were a negative number, it would change the direction of the inequality.) Since $x > z$, then

$$\frac{1}{x} < \frac{1}{z} \text{ and } \frac{1}{xy} < \frac{1}{yz}.$$

The correct choice is (B).

• **PROBLEM 2-88**

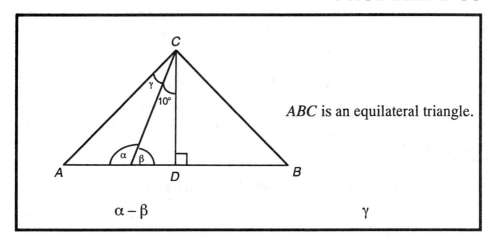

*ABC* is an equilateral triangle.

$$\alpha - \beta \qquad\qquad \gamma$$

## SOLUTION:

**(C)**    *ABC* is an equilateral triangle.

$\angle CAB = \angle ABC = \angle ACB = 60°$

Since $\angle ADC = \angle CDB = 90°$, and since the triangle is equilateral, $\overline{CD}$ bisects $\angle ACB$.

Therefore, $\angle ACD = 30°$

$\gamma + 10° = 30°$

$\gamma = 20°$

In $\triangle CDE$ we have $10° + 90° + \beta = 180°$, because the sum of all the angles in a triangle is $180°$. Solving for $\beta$, we find $\beta = 80°$.

Since $\alpha$ and $\beta$ are supplementary, then

$\alpha + \beta = 180°$

$\alpha + 80° = 180°$

$\alpha = 100°$

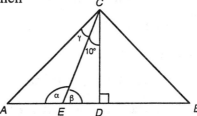

So    $\alpha - \beta = 100° - 80° = 20°$.

Also, we found earlier that $\gamma = 20°$.

Hence, Column A equals Column B.

The correct choice is (C).

## • PROBLEM 2-89

Given two numbers whose sum is 23
and whose product is 120

| The difference of the two numbers | The smaller of the two numbers |

### SOLUTION:

**(B)** Let $a$ be the first number and $b$ be the second number.

$$a + b = 23 \qquad\qquad (1)$$

$$ab = 120$$

We must solve for $a$ in terms of $b$. We can then substitute this value for $a$ and solve for $b$.

$$a = 23 - b$$

$$ab = 120$$

$$(23 - b)\, b = 120$$

$$23b - b^2 = 120$$

$$b^2 - 23b + 120 = 0$$

This factors to

$$(b - 8)(b - 15) = 0$$

$$b = 8 \text{ or } 15$$

From equation (1), when

$$b = 8, a = 15,$$

and when $\quad b = 15, a = 8$.

Hence, the two numbers are 8 and 15.

Thus, the difference between the two numbers is 7. The smaller of the two numbers is 8. So, column B is larger.

The correct choice is (B).

## • PROBLEM 2-90

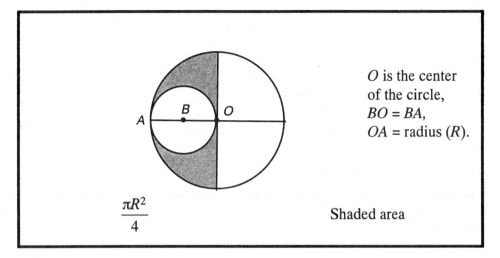

$O$ is the center of the circle, $BO = BA$, $OA$ = radius ($R$).

$$\frac{\pi R^2}{4}$$

Shaded area

## SOLUTION:

**(C)**

$OA = R$ = radious of large circle

$BO = BA$

$BA = \dfrac{R}{2}$ = radius of small circle

shaded area = area of half of circle with radius $R$

– area of circle with radius $\dfrac{R}{2}$

$$= \frac{1}{2}\pi R^2 - \pi\left(\frac{R}{2}\right)^2$$

$$= \frac{\pi R^2}{2} - \frac{\pi R^2}{4}$$

$$= \frac{2\pi R^2}{4} - \frac{\pi R^2}{4}$$

$$= \frac{\pi R^2}{4}$$

Therefore, Column A equals Column B.

The correct choice is (B).

• **PROBLEM 2-91**

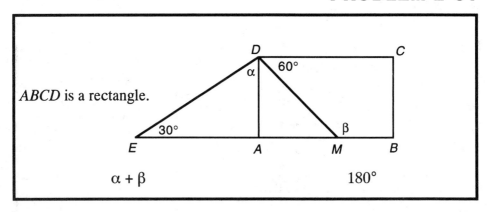

*ABCD* is a rectangle.

$\alpha + \beta$                  180°

## SOLUTION:

**(C)**     *ABCD* is a rectangle.

In the triangle *EAD* we see that

$$90° + 30° + \alpha = 180°$$
$$\alpha = 180° - 120° = 60°$$

In the quadrilateral *DMBC*, we can solve for $\beta$ as follows:

$$60° + \beta + 90° + 90° = 360°.$$

This is because the sum of the angles of a quadrilateral is 360°. Solving for $\beta$ we get

$$\beta = 360° - 60° - 90° - 90°$$
$$= 360° - 240°$$
$$\beta = 120°$$

Therefore $\alpha + \beta = 60° + 120° = 180°$.

The correct choice is (C).

## • PROBLEM 2-92

Given two squares, with the area of the second
four times the area of the first square

Four times the perimeter of          The perimeter of
the first square                     the second square

### SOLUTION:

(A)    Let

$x$ = the length of the side of the first square

$y$ = the length of the side of the second square

We are given that the area of the second square is equal to four times the area of the first square. This can be written as follows:

$y^2 = 4x^2$.

Simplifying $y = 2x$.

The perimeter of the first square is

$4 \times x = 4x$.

The perimeter of the second square is

$4y = 4(2x) = 8x$.

Thus, we see that the perimeter of the second square is twice the perimeter of the first square. Therefore, four times the perimeter of the first square is larger than the perimeter of the second square.

The correct choice is (A).

## • PROBLEM 2-93

$$x < 0$$
$$y < 0$$

$x - y$                           0

## SOLUTION:

**(D)**

$$x < 0$$

$$y < 0$$

If $x = -3$, $y = -4$, then

$$x - y = (-3) - (-4) = -3 + 4 = 1.$$

If $x = -4$, $y = -3$, then

$$x - y = (-4) - (-3) = -4 + 3 = -1.$$

Therefore, we cannot infer any relationship between Column A and Column B.

The correct choice is (D).

### • PROBLEM 2–94

$$m < 0, n < 0, \text{ and } m > n.$$

$$\frac{m}{n} \qquad\qquad\qquad \frac{n}{m}$$

## SOLUTION:

**(B)**

$$m < 0, n < 0, m > n$$

Since both $m$ and $n$ arenegative,

$$\frac{m}{n} \text{ and } \frac{n}{m}$$

are both positive numbers (minus signs in numerator and denominator cancel).

Since $m > n$,

$$\frac{m}{n} < \frac{n}{n}$$

Try an example:

$$m = -5$$

$$n = -10$$

(−10 is to the left of −5 on the number line, so −10 is smaller than −5.)

$$\frac{m}{n} = \frac{-5}{-10} = \frac{1}{2}$$

$$\frac{n}{m} = \frac{-10}{-5} = 2$$

So,  $\dfrac{m}{n} < \dfrac{n}{m}.$

If you try fractions such as $m = -1/4$ and $n = -1/2$, you will again find that

$$\frac{m}{n} < \frac{n}{m}.$$

The correct choice is (B).

## • PROBLEM 2-95

In a Boeing 747 with 320 seats, 73 women are traveling and 20% of the seats are empty.

Men that are traveling                                    185

## SOLUTION:

**(B)**    Let $x$ = the number of men traveling on the Boeing 747.

If 20% of the total seats are empty, then 20% of 320 are empty.

20% of 320 = .20 × 320 = 64 empty seats

The number of men traveling + the number of women traveling + the number of empty seats = 320

$$x + 73 + 64 = 320$$

$$x = 320 - 73 - 64$$

$$x = 183$$

Column A = 183 and Column B = 185.

Hence, the correct choice is (B).

## • PROBLEM 2-96

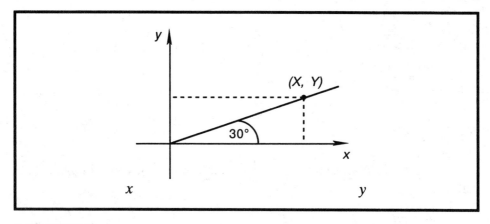

## SOLUTION:

**(A)**    The side opposite the smaller angle (30°) must be the shorter side.

Thus, the side labelled $y$ must be shorter than the side labelled $x$.

Since the triangle is located in the first quadrant, both $x$ and $y$ are positive, and $x > y$.'

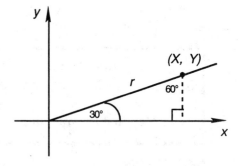

If the triangle were located in a different quadrant, you would attach signs appropriate to that quadrant, and then decide whether $x$ or $y$ was larger, taking the signs into account.

Be sure to note whether a given problem is asking for the length of the sides (length is defined to be positive, so signs don't matter), or asking for $x$ and $y$ coordinates, where signs do matter.

The correct answer is (A).

## • PROBLEM 2-97

$$\underbrace{3 + 3 + 3 + \ldots}_{m \text{ times}} > \underbrace{4 + 4 + 4 + 4 + \ldots}_{p \text{ times}}$$

$$\frac{m}{p} \qquad\qquad \frac{4}{3}$$

### SOLUTION:

**(A)**    We can rewrite $3 + 3 + 3 + \ldots$ *m* times as $3m$. Likewise, $4 + 4 + 4 + 4 + \ldots$ *p* times can be written as $4p$.

$3m > 4p$ is a given statement.

*m* and *p* are *positive*, so we can divide both sides by *p* without changing the direction of the inequality.

$$\frac{m}{p} > \frac{4}{3}$$

Hence, Column A is larger than Column B.

The correct choice is (A).

## • PROBLEM 2-98

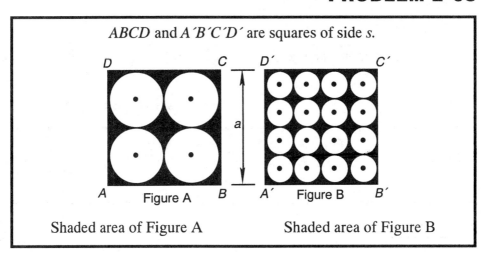

*ABCD* and *A´B´C´D´* are squares of side *s*.

Shaded area of Figure A          Shaded area of Figure B

## SOLUTION:

**(C)**    If *ABCD* is a square with side *s*, then each circle has a radius of $s/4$.

The shaded area = area of the square − area of the four circles.

$$\text{Shaded area} = s^2 - 4\left[\pi\left(\frac{s}{4}\right)^2\right]$$

$$= s^2 - \frac{4\pi s^2}{16}$$

$$= s^2 - \frac{\pi s^2}{4}$$

$$= s^2\left(1 - \frac{\pi}{4}\right)$$

In *A'B'C'D'* each circle has a radius of $s/8$. There are 16 circles.

$$\text{Shaded area} = s^2 - 16\left[\pi\left(\frac{s}{8}\right)^2\right]$$

$$= s^2 - \frac{16\pi s^2}{64}$$

$$= s^2\left(1 - \frac{\pi}{4}\right)$$

The shaded areas are equal.

The correct choice is (C).

## • PROBLEM 2-99

For $y \neq 0$, let $\boxed{y} = (y^2 - 1)/y$.

$\boxed{-3}$                                      $-3$

## SOLUTION:

**(A)**

$y \neq 0$

$$\boxed{y} = \frac{(y^2 - 1)}{y}$$

For $\boxed{-3} = \frac{((-3)^2 - 1)}{-3} = \frac{9-1}{-3} = -\frac{8}{3}$

$\boxed{-3} = -\frac{8}{3} > -3$

Hence, the correct choice is (A).

## • PROBLEM 2-100

## SOLUTION:

**(C)**

$$\boxed{1} = \boxed{y} = \frac{(y^2 - 1)}{y} = \frac{(1)^2 - 1}{1} = \frac{1-1}{1} = \frac{0}{1} = 0$$

Similarly,

$$\boxed{-1} = \frac{((-1)^2 - 1)}{-1} = \frac{1-1}{-1} = \frac{0}{-1} = 0$$

Therefore, the correct choice is (C).

# Chapter 3
# Student-Produced Response

# CHAPTER 3

# STUDENT-PRODUCED RESPONSE

The Student-Produced Response format of the SAT I is designed to give the student a certain amount of flexibility in answering questions. In this section the student must calculate the answer to a given question and then enter the solution into a grid. The grid is constructed so that a solution can be given in either decimal or fraction form. Either form is acceptable unless otherwise stated.

The problems in the Student-Produced Response section try to reflect situations arising in the real world. Here calculations will involve objects occurring in everyday life. There is also an emphasis on problems involving data interpretation. In keeping with this emphasis, students will be allowed the use of a calculator during the exam.

Through this review, you will learn how to successfully attack Student-Produced Response questions. Familiarity with the test format combined with solid math strategies will prove invaluable in answering the questions quickly and accurately.

## ABOUT THE DIRECTIONS

Each Student-Produced Response question will require you to solve the problem and enter your answer in a grid. There are specific rules you will need to know for entering your solution. If you enter your answer in an incorrect form, you will not receive credit, even if you originally solved the problem correctly. Therefore, you should carefully study the following rules now, so you don't have to waste valuable time during the actual test:

**DIRECTIONS**: Each of the following questions requires you to solve the problem and enter your answer in the ovals in the special grid:

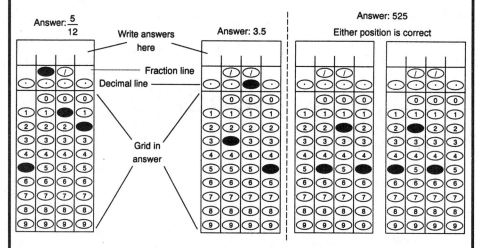

- You may begin filling in your answer in any column, space permitting. Columns not needed should be left blank.

- Answers may be entered in either decimal or fraction form. For example, $^3/_{12}$ or .25 are equally acceptable.

- A mixed number, such as $4^1/_2$, must be entered either as 4.5 or $^9/_2$. If you entered 41/2, the machine would interpret your answer as $^{41}/_2$, not $4^1/_2$.

- There may be some instances where there is more than one correct answer to a problem. Should this be the case, grid only one answer.

- Be very careful when filling in ovals. Do not fill in more than one oval in any column, and make sure to completely darken the ovals.

- It is suggested that you fill in your answer in the boxes above each column. Although you will not be graded incorrectly if you do not write in your answer, it will help you fill in the corresponding ovals.

- If your answer is a decimal, grid the most accurate value possible. For example, if you need to grid a repeating decimal such as $0.66\overline{66}$, enter the answer as .666 or .667. A less accurate value, such as .66 or .67, is not acceptable.

- A negative answer cannot appear for any question.

- Ignore any dollar signs or percentage symbols when gridding your answer.

### SAMPLE QUESTION

How many pounds of apples can be bought with $5.00 if apples cost $.40 a pound?

### SOLUTION

Converting dollars to cents we obtain the equation

$x = 500 \div 40$

$x = 12.5$

The solution to this problem would be gridded as

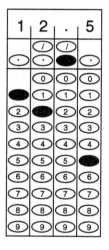

# ABOUT THE QUESTIONS

Within the SAT I Student-Produced Response section you will be given 10 questions. You will have 30 minutes to answer these questions in addition to 15 Quantitative Comparison questions. This means you will be required to answer 25 questions in 30 minutes. Therefore, you should work quickly.

The Student-Produced Response questions will come from the areas of arithmetic, algebra, and geometry. There is an emphasis on word problems and on data interpretation, which usually involves reading tables to answer questions. Many of the geometry questions will refer to diagrams or will ask you to create a figure from information given in the question.

The following will detail the different types of questions you should expect to encounter in the Student-Produced Response section.

# ARITHMETIC QUESTIONS

These arithmetic questions test your ability to perform standard manipulations and simplifications of arithmetic expressions. For some questions, there is more than one approach. There are six kinds of arithmetic questions you may encounter in the Student-Produced Response section. For each type of question, we will show how to solve the problem and grid your answer.

## Question Type 1: Properties of a Whole Number *N*

This problem tests your ability to find a whole number with a given set of properties. You will be given a list of properties of a whole number and asked to find that number.

*PROBLEM*

> The properties of a whole number $N$ are
>
> (A)  $N$ is a perfect square.
>
> (B)  $N$ is divisible by 2.
>
> (C)  $N$ is divisible by 3.
>
> Grid in the second smallest whole number with the above properties.

*SOLUTION*

Try to first obtain the smallest number with the above properties. The smallest number with properties (B) and (C) is 6. Since property (A) says the number must be a perfect square, the smallest number with properties (A), (B), and (C) is 36.

$6^2 = 36$ is the smallest whole number with the above properties. The second smallest whole number (the solution) is

$$2^2 6^2 = 144.$$

The correct answer entered into the grid is

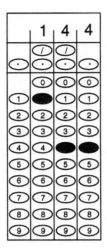

## Question Type 2: Simplifying Fractions

This type of question requires you to simplify fractional expressions and grid the answer in the format specified. By canceling out terms common to both the numerator and denominator, we can simplify complex fractional expressions.

### *PROBLEM*

Change $\dfrac{1}{2} \times \dfrac{3}{7} \times \dfrac{2}{8} \times \dfrac{14}{10} \times \dfrac{1}{3}$ to decimal form.

### *SOLUTION*

The point here is to cancel out terms common to both the numerator and denominator. Once the fraction is brought down to lowest terms, the result is entered into the grid as a decimal.

After cancellation we are left with the fraction $^1/_{40}$. Equivalently,

$$\frac{1}{40} = \frac{1}{10} \times \frac{1}{4} = \frac{1}{10}(.25) = .025.$$

Hence, in our grid we enter

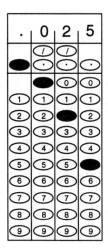

Note: If "decimal" was not specified, any correct version of the answer could be entered into the grid.

## Question Type 3: Prime Numbers of a Particular Form

Here, you will be asked to find a prime number with certain characteristics. Remember—a prime number is a number that can only be divided by itself and 1.

*PROBLEM*

Find a prime number of the form $7k + 1 < 50$.

*SOLUTION*

This is simply a counting problem. The key is to list all the numbers of the form $7k + 1$ starting with $k = 0$. The first one that is prime is the solution to the problem.

The whole numbers of the form $7k + 1$ which are less than 50 are 1, 8, 15, 22, 29, 36, and 43. Of these, 29 and 43 are prime numbers. The possible solutions are 29 and 43.

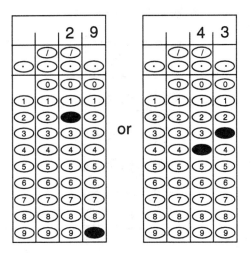

## Question Type 4: Order of Operations

The following question type tests your knowledge of the arithmetic order of operations. Always work within the parentheses or with absolute values first, while keeping in mind that multiplication and division are carried out before addition and subtraction.

### PROBLEM

Find a solution to the equation $x \div 3 \times 4 \div 2 = 6$.

### SOLUTION

The key here is to recall the order of precedence for arithmetic operations. After simplifying the expression one can solve for $x$.

Since multiplication and division have the same level of precedence, we simplify the equation from left to right to obtain

$$\frac{x}{3} \times 4 \div 2 = 6$$

$$\frac{4x}{3} \div 2 = 6$$

$$\frac{2x}{3} = 6$$

$$x = 9$$

As 9 solves the above problem, our entry in the grid is

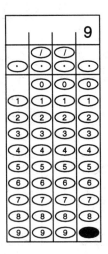

## Question Type 5: Solving for Ratios

This type of question tests your ability to manipulate ratios given a set of constraints.

### PROBLEM

Let $A$, $B$, $C$, and $D$ be positive integers. Assume that the ratio of $A$ to $B$ is equal to the ratio of $C$ to $D$. Find a possible value for $A$ if the product of $BC = 24$ and $D$ is odd.

### SOLUTION

The quickest way to find a solution is to list the possible factorizations of 24:

$1 \times 24$

$2 \times 12$

$3 \times 8$

$4 \times 6$

Since $AD = BC = 24$ and $D$ is odd, the only possible solution is $A = 8$ (corresponding to $D = 3$).

In the following grid we enter

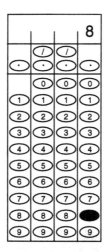

## Question Type 6: Simplifying Arithmetic Expressions

Here you will be given an arithmetic problem that is easier to solve if you transform it into a basic algebra problem. This strategy saves valuable time by cutting down on the number and complexity of computations involved.

*PROBLEM*

Simplify $1 - \left(\dfrac{1}{2} + \dfrac{1}{4} + \dfrac{1}{8} + \dfrac{1}{16} + \dfrac{1}{32} + \dfrac{1}{64}\right)$.

*SOLUTION*

This problem can be done one of two ways. The "brute force" approach would be to get a common denominator and simplify. An approach involving less computation is given below.

$$\text{Set } S = 1 - \left(\frac{1}{2} + \frac{1}{4} + \frac{1}{8} + \frac{1}{16} + \frac{1}{32} + \frac{1}{64}\right)$$

Multiplying this equation by 2 we obtain

$$2S = 2 - \left(1 + \frac{1}{2} + \frac{1}{4} + \frac{1}{8} + \frac{1}{16} + \frac{1}{32}\right)$$

$$2S = 1 - \left(\frac{1}{2} + \frac{1}{4} + \frac{1}{8} + \frac{1}{16} + \frac{1}{32}\right)$$

$$2S = 1 - \left(\frac{1}{2} + \frac{1}{4} + \frac{1}{8} + \frac{1}{16} + \frac{1}{32} + \frac{1}{64}\right) + \frac{1}{64}$$

$$2S = S + \frac{1}{64}$$

$$S = \frac{1}{64}$$

We enter into the grid

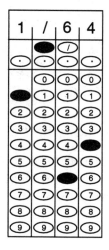

# ALGEBRA QUESTIONS

Within the Student-Produced Response section, you will also encounter algebra questions which will test your ability to solve algebraic expressions in the setting of word problems. You may encounter the following six types of algebra questions during the SAT I. As in the previous section, we provide methods for approaching each type of problem.

### Question Type 1: Solving a System of Linear Equations

This is a standard question which will ask you to find the solution to a system of two linear equations with two unknowns.

*PROBLEM*

Consider the system of simultaneous equations given by

$$y - 2 = x - 4$$

$$y + 3 = 6 - x$$

Solve for the quantity $6y + 3$.

**SOLUTION**

This problem can be solved by taking the first equation given and solving for $x$. This would yield

$$x = y + 2.$$

Next, we plug this value for $x$ into the second equation, giving us

$$y + 3 = 6 - (y + 2).$$

Solve this equation for $y$ and we get

$$y = \frac{1}{2}.$$

We are asked to solve for $6y + 3$, so we can plug our value for $y$ in and get

$$6\left(\frac{1}{2}\right) + 3 = 6.$$

Our answer is 6 and gridded correctly it is

## Question Type 2: Word Problems Involving Age

When dealing with this type of question, you will be asked to solve for the age of a particular person. The question may require you to determine how much older one person is, how much younger one person is, or the specific age of the person.

**PROBLEM**

Tim is 2 years older than Jane and Joe is 4 years younger than Jane. If the sum of the ages of Jane, Joe, and Tim is 28, how old is Joe?

**SOLUTION**

Define Jane's age to be the variable $x$ and work from there.

Let

Jane's age = $x$

Tim's age = $x + 2$

Joe's age = $x - 4$

Summing up the ages we get

$$x + x + 2 + x - 4 = 28$$

$$3x - 2 = 28$$

$$3x = 30$$

$$x = 10$$

Joe's age = $10 - 4 = 6$.

Hence, we enter into the grid

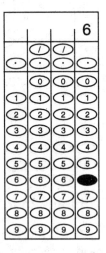

## Question Type 3: Word Problems Involving Money

Word problems involving money will test your ability to translate the information given into an algebraic statement. You will also be required to solve your algebraic statement.

**PROBLEM**

After receiving his weekly paycheck on Friday, a man buys a television for $100, a suit for $200, and a radio for $50. If the total money he spent amounts to 40% of his paycheck, what is his weekly salary?

**SOLUTION**

Simply set up an equation involving the man's expenditures and the percentage of his paycheck that he used to buy them.

Let the amount of the man's paycheck equal $x$. We then have the equation

$$40\% x = 100 + 200 + 50$$

$$0.4x = 350$$

$$x = \$875$$

In the grid we enter

## Question Type 4: Systems of Non-Linear Equations

This type of question will test your ability to perform the correct algebraic operations for a given set of equations in order to find the desired quantity.

**PROBLEM**

> Consider the system of equations
>
> $$x^2 + y^2 = 8$$
>
> $$xy = 4$$
>
> Solve for the quantity $3x + 3y$.

**SOLUTION**

Solve for the quantity $x + y$ and not for $x$ or $y$ individually.

First, multiply the equation $xy = 4$ by 2 to get $2xy = 8$. Adding this to $x^2 + y^2 = 8$ we obtain

$$x^2 + 2xy + y^2 = 16$$

$$(x + y)^2 = 16$$

$$x + y = 4$$

or

$$x + y = -4$$

Hence, $3x + 3y = 12$ or $3x + 3y = -12$. We enter 12 for a solution since $-12$ cannot be entered into the grid.

## Question Type 5: Word Problems Involving Hourly Wage

When dealing with this type of question, you will be required to form an algebraic expression from the information based on a person's wages. You will then solve the expression to determine the person's wages (i.e., hourly, daily, annually, etc.).

## PROBLEM

Jim works 25 hours a week earning $10 an hour. Sally works 50 hours a week earning $y$ dollars an hour. If their combined income every two weeks is $2,000, find the amount of money Sally makes an hour.

## SOLUTION

Be careful. The combined income is given over a two-week period.

Simply set up an equation involving income. We obtain

$$2[(25)(10) + (50)(y)] = 2,000$$

$$[(25)(10) + (50)(y)] = 1,000$$

$$250 + 50y = 1,000$$

$$50y = 750$$

$$y = \$15 \text{ an hour}$$

We enter in the grid

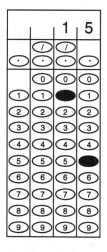

## Question Type 6: Word Problems Involving Consecutive Integers

In this type of question, you will need to set up an equation involving consecutive integers based on the product of the integers, which is given.

## PROBLEM

Consider two positive consecutive odd integers such that their product is 143. Find their sum.

## SOLUTION

Be careful. Notice $x$ and $y$ are consecutive odd integers.

Let

1st odd integer $= x$

2nd odd integer $= x + 2$

We get

$$x(x + 2) = 143$$
$$x^2 + 2x - 143 = 0$$
$$(x - 11)(x + 13) = 0$$

Hence $\qquad x = 11$

and $\qquad x = -13.$

From the above we obtain the solution sets $\{11, 13\}$ and $\{-13, -11\}$ whose sums are 24 and $-24$, respectively. Since the problem specifies that the integers are positive, we enter 24.

# GEOMETRY QUESTIONS

In this section, we will explain how to solve questions which test your ability to find the area of various geometric figures. There are six types of questions you may encounter.

## Question Type 1: Area of an Inscribed Triangle

This question asks you to find the area of a triangle which is inscribed in a square. By knowing certain properties of both triangles and squares, we can deduce the necessary information.

*PROBLEM*

Consider the triangle inscribed in the square.

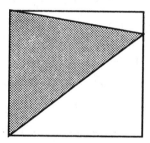

If the area of the square is 36, find the area of the triangle.

*SOLUTION*

Find the height of the triangle.

Let $x$ be the length of the square. Since the four sides of a square are equal, and the area of a square is the length of a side squared, $x^2 = 36$. Therefore, $x = 6$.

The area of a triangle is given by

$\frac{1}{2}$ (base) (height).

Here $x$ is both the base and height of the triangle. The area of the triangle is

$\frac{1}{2}$ (6) (6) = 18.

This is how the answer would be gridded.

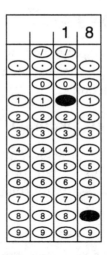

## Question Type 2: Length of the Side of a Triangle

For this type of question, one must find the length of a right triangle given information about the other sides. The key here is to apply the Pythagorean Theorem, which states that the square of the hypotenuse of a right triangle is equal to the sum of the squares of the other two sides.

### PROBLEM

Consider the line given below

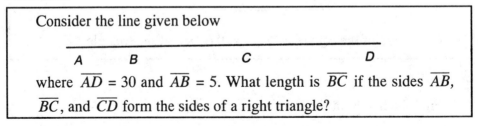

where $\overline{AD}$ = 30 and $\overline{AB}$ = 5. What length is $\overline{BC}$ if the sides $\overline{AB}$, $\overline{BC}$, and $\overline{CD}$ form the sides of a right triangle?

### SOLUTION

Draw a diagram and fill in the known information.

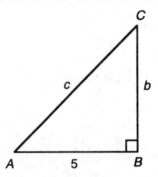

Next, apply the Pythagorean Theorem ($a^2 + b^2 = c^2$), filling in the known variables. Here, we are solving for $\overline{BC}$ ($b$ in our equation). We know that

$a = 5$, and since $\overline{AD} = 30$ and $\overline{AB} = 5$, $\overline{BD} = 25$. Filling in these values, we obtain this equation:

$$5^2 + x^2 = (25 - x)^2$$
$$25 + x^2 = 625 - 50x + x^2$$
$$50x = 600$$
$$x = 12$$

This is one possible solution. If we had chosen $x = \overline{CD}$ and $25 - x = \overline{BC}$, one obtains $\overline{BC} = 13$, which is another possible solution. The possible grid entries are shown here.

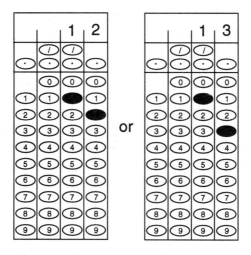

## Question Type 3: Solving for the Degree of an Angle

Here you will be given a figure with certain information provided. You will need to deduce the measure of an angle based both on this information, as well as other geometric principles. The easiest way to do this is by setting up an algebraic expression.

### PROBLEM

Find the measure of the angle $y$ in the diagram below.

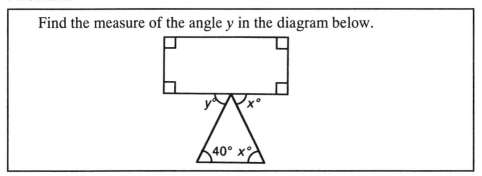

## SOLUTION

Use the fact that the sum of the angles on the bottom side of the box is 180°.

Let $z$ be the angle at the top of the triangle. Since we know the sum of the angles of a triangle is 180°,

$z = 180 - (x + 40)$.

Summing all the angles at the bottom of the square, we get

$$y + [180 - (x + 40)] + x = 180$$

$$y + 140 - x + x = 180$$

$$y + 140 = 180$$

$$y = 40$$

In the grid we enter

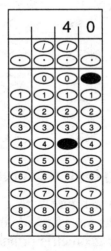

## Question Type 4: Solving for the Length of a Side

For this type of question, you will be given a figure with certain measures of sides filled in. You will need to apply geometric principles to find the missing side.

**PROBLEM**

Consider the figure below.

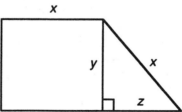

In the figure let $x$ and $y$ be whole numbers where $xy = 65$. Also assume the area of the whole figure is 95 square inches. Find $y$.

**SOLUTION**

The key point here is that $x$ and $y$ are whole numbers. Using the figure we only have a finite number of possibilities for $z$.

The equation for the area of the above figure is

$$xy + yz = 95.$$

Substituting $xy = 65$ into the above equation, we get

$$yz = 30.$$

Using the fact that $xy = 65$, we know $y$ can be either 1, 5, or 13. As $y = 13$ does not yield a factorization for $yz = 60$, $y$ is either 1 or 5. If $y = 1$, this implies $x = 65$ and $z = 60$ which contradicts the Pythagorean Theorem (i.e., $1^2 + 60^2 = 13^2$). If $y = 5$, this implies $x = 13$ and $z = 12$ which satisfies $y^2 + z^2 = x^2$; hence, the solution is $y = 5$.

In our grid we enter

## Question Type 5: Solving for the Area of a Region

Here, you will be given a figure with a shaded region. Given certain information, you will need to solve for the area of that region.

### PROBLEM

Consider the concentric squares drawn below.

Assume that the side of the larger square is length 1. Also assume that the smaller square's perimeter is equal to the diagonal of the larger square. Find the area of the shaded region.

### SOLUTION

The key here is to find the length of the side for the smaller square.

By the Pythagorean Theorem the diameter of the square is

$$d^2 = 1^2 + 1^2$$

which yields $d = \sqrt{2}$. Similarly, the smaller square's perimeter is $\sqrt{2}$; hence, the smaller square's side

$$= \frac{\sqrt{2}}{4}.$$

Calculating the area for the shaded region, we get

$$A = A_{\text{large}} - A_{\text{small}}$$

$$A = 1 - \left(\frac{\sqrt{2}}{4}\right)^2$$

$$A = 1 - \frac{2}{16}$$

$$A = \frac{7}{8}$$

In the grid we enter

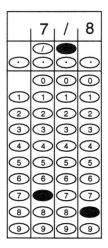

## Question Type 6: Solve for a Sum of Lengths

The question here involves solving for a sum of lengths in the figure given knowledge pertaining to its area.

*PROBLEM*

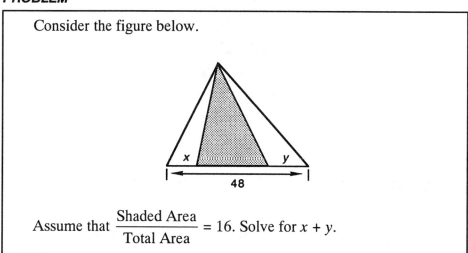

Consider the figure below.

Assume that $\dfrac{\text{Shaded Area}}{\text{Total Area}} = 16$. Solve for $x + y$.

*SOLUTION*

Solve for $x + y$ and not for $x$ or $y$ individually. Denote by $b$ the base of the smaller triangle. Then

$$48 - b = x + y.$$

From the information given

$$\frac{\frac{1}{2}48h}{\frac{1}{2}bh} = 16$$

$$\frac{1}{2}48h = 16\left(\frac{1}{2}bh\right)$$

$$24 = 8b$$

$$3 = b$$

This yields

$$x + y = 48 - 3 = 45.$$

We enter in the grid

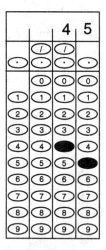

# ANSWERING STUDENT-PRODUCED RESPONSE QUESTIONS

When answering Student-Produced Response questions, you should follow these steps.

| STEP 1 | Identify the type of question with which you are presented (i.e., arithmetic, algebra, or geometry). |

| STEP 2 | Once you have determined if the question deals with arithmetic, algebra, or geometry, further classify the question. Then, try to determine what type of arithmetic (or algebra or geometry) question is being presented. |

STEP 3 | Solve the question using the techniques explained in this review. Make sure your answer can be gridded.

STEP 4 | Grid your answer in the question's corresponding answer grid. Make sure you are filling in the correct grid. Keep in mind that it is not mandatory to begin gridding your answer on any particular side of the grid. Fill in the ovals as completely as possible, and beware of any stray marks—stray lines may cause your answer to be marked incorrect.

The drill questions which follow should be completed to help reinforce the material which you have just studied. Be sure to refer back to the review if you need help answering the questions.

**DIRECTIONS**: Each of the following questions requires you to solve the problem and enter your answer in the ovals in the special grid:

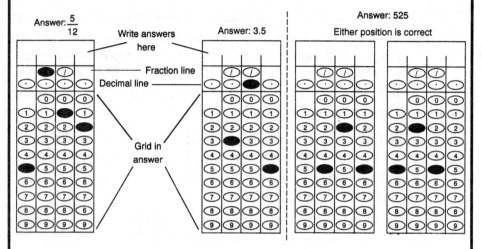

- You may begin filling in your answer in any column, space permitting. Columns not needed should be left blank.

- Answers may be entered in either decimal or fraction form. For example, $^3/_{12}$ or .25 are equally acceptable.

- A mixed number, such as $4^1/_2$, must be entered either as 4.5 or $^9/_2$. If you entered 41/2, the machine would interpret your answer as $^{41}/_2$, not $4^1/_2$.

- There may be some instances where there is more than one correct answer to a problem. Should this be the case, grid only one answer.

- Be very careful when filling in ovals. Do not fill in more than one oval in any column, and make sure to completely darken the ovals.

- It is suggested that you fill in your answer in the boxes above each column. Although you will not be graded incorrectly if you do not write in your answer, it will help you fill in the corresponding ovals.

- If your answer is a decimal, grid the most accurate value possible. For example, if you need to grid a repeating decimal such as 0.6666, enter the answer as .666 or .667. A less accurate value, such as .66 or .67, is not acceptable.

- A negative answer cannot appear for any question.

- Ignore any dollar signs or percentage symbols when gridding your answer.

### • PROBLEM 3–1

What does $(-5 - 5) \times (-5 - (-5))$ equal?

## SOLUTION:

The correct response is:

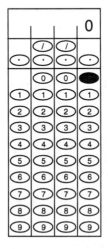

$$(-5 - 5) \times (-5 - (-5)) = x$$
$$(-5 - 5) \times (-5 + 5) = x$$
$$(-10) \times (0) = x$$
$$0 = x$$

## • PROBLEM 3-2

After 5 tests had been given in a student's biology class, the student had an 82 average (arithmetic mean). After one more test, the student's average was 84. What grade did the student receive on the sixth test?

## SOLUTION:

The correct response is:

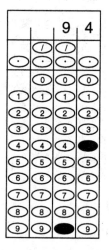

If the student had an average of 82 after 5 tests, then the student accumulated

$82 \times 5 = 410$ points for 5 tests.

After one more test, the student's average was 84. Thus, the student accumulated

$84 \times 6 = 504$ points for 6 tests.

This means that the student would have scored $504 - 410 = 94$ points on the last exam.

## • PROBLEM 3–3

At the end of the month, a woman pays $714 in rent. If the rent constitutes 21% of her monthly income, what is her hourly wage given the fact that she works 34 hours per week, and that there are 4 weeks per month?

## SOLUTION:

The correct response is:

If the rent constitutes 21% of the monthly income, then let $x$ = monthly income and $R$ = rent = $714.

$$R = .21\, x = \$714$$

$$x = \frac{714}{.21} = \$3,400$$

Assuming that her monthly income is $3,400 and that there are 4 weeks in one month, we can calculate her weekly income.

weekly income = $3,400/month × month/4 weeks

= 3,400/4 weeks

= $850/week

Since the woman works 34 hours per week, her hourly wage can be computed as follows:

$850/week × 1week/34 hours = $850/34 hours

= $25/hour

## • PROBLEM 3-4

Find the largest integer which is less than 100 and divisible by 3 and 7.

### SOLUTION:

The correct response is:

Numbers that are divisible by both 3 and 7 are also divisible by 21. We want the largest integer multiple of 21 that is smaller than 100. Multiples of 21 are 21, 42, 63, 84, 105.

84 is the largest multiple that is also less than 100.

## • PROBLEM 3-5

The radius of the smaller of two concentric circles is 5 cm, while the radius of the larger circle is 7 cm. Determine the area of the shaded region.

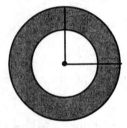

## SOLUTION:

The correct response is:

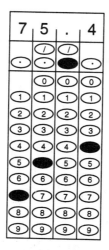

The area of the shaded region is equal to the area of the larger circle minus the area of the smaller circle.

area of circle with radius of 7 – area of circle with radius of 5

$\pi r^2$ $\qquad\qquad\qquad\qquad$ $\pi r^2$

$\pi(7)^2$ $\qquad\qquad\qquad\qquad$ $\pi(5)^2$

$49\pi$ $\qquad\qquad\qquad\qquad$ $25\pi$

$49\pi - 25\pi = 24\pi$

Since $\pi = 3.14$,

$24\pi = 24\,(3.14) = 75.36$.

### • PROBLEM 3-6

What is the value of the following?

$$\frac{1}{6} + \frac{2}{3} + \frac{1}{6} - \frac{1}{3} + 1 - \frac{3}{4} - \frac{1}{4} =$$

## SOLUTION:

The correct response is:

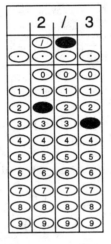

We can simplify this expression by finding a least common denominator (LCD). We see that 12 is the LCD. We then add the numbers from left to right.

$$\frac{1}{6} \times \left(\frac{2}{2}\right) + \frac{2}{3} \times \left(\frac{4}{4}\right) + \frac{1}{6} \times \left(\frac{2}{2}\right) - \frac{1}{3}\left(\frac{4}{4}\right) + 1\left(\frac{12}{12}\right) - \frac{3}{4}\left(\frac{3}{4}\right) - \frac{1}{4}\left(\frac{3}{3}\right) =$$

$$= \frac{2}{12} + \frac{8}{12} + \frac{2}{12} - \frac{4}{12} + \frac{12}{12} - \frac{9}{12} - \frac{3}{12}$$

$$= \frac{2 + 8 + 2 - 4 + 12 - 9 - 3}{12}$$

$$= \frac{8}{12}$$

$$= \frac{2}{3}$$

### • PROBLEM 3-7

Find *x*.

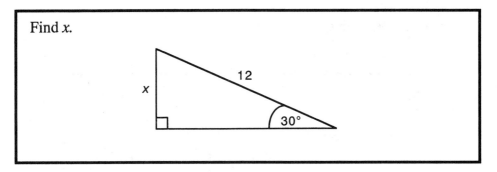

## SOLUTION:

The correct response is:

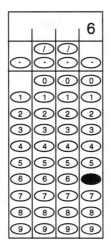

This problem can be solved two different ways.

I.    $\sin (30°) = (.500) = \dfrac{x}{12}$

We needed to look up the sin of 30°.

$x = (.500)(12)$

$x = 6$

II.    The triangle is a 30° – 60° – 90° triangle. Therefore, the side opposite the 30° angle is equal to half the hypotenuse.

$x = \dfrac{1}{2}$ hypotenuse

$x = \dfrac{1}{2}(12) = 6$

**• PROBLEM 3-8**

Evaluate the following expression.

$$|-8-4| \div 3 \times 6 + (-4) =$$

## SOLUTION:

The correct response is:

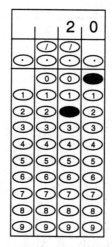

$$|-8-4| \div 3 \times 6 + (-4) = |-12| \div 3 \times 6 + (-4)$$

$$= 12 \div 3 \times 6 + (-4)$$

$$= 4 \times 6 + (-4)$$

$$= 24 + (-4) = 20$$

## • PROBLEM 3-9

Let $\overline{RO}$ = 16, $\overline{HM}$ = 30. Find the perimeter of rhombus *HOMR*.

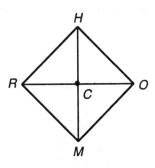

## SOLUTION:

The correct response is:

Taking the square roots of both sides we have $\overline{RH}$.

$$289 = (\overline{RH})^2$$

$$17 = \overline{RH}$$

Thus, the side of the rhombus has a length = 17. Since a rhombus has four equal sides, the perimeter is 4 (17) = 68.

### • PROBLEM 3-10

Six years ago, Henry's mother was nine times as old as Henry. Now she is only three times as old as Henry. How old is Henry now?

## SOLUTION:

The correct response is:

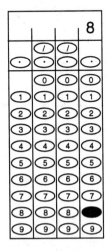

Let $x$ = Henry's age now

$3x$ = age of Henry's mother now

$x - 6$ = Henry's age six years ago

$3x - 6$ = age of Henry's mother six years ago

$$3x - 6 = 9 (x - 6)$$

$$3x - 6 = 9x - 54$$

$$3x - 6 + 54 = 9x$$

$$3x + 48 = 9x$$

$$48 = 6x$$

$$\frac{48}{6} = x$$

$$8 = x$$

Henry is now 8 years old.

## • PROBLEM 3-11

If tangerines are sold at the rate of 12 for $1.60 and Sally buys 3 tangerines with a $10.00 bill, and if there is no tax on this transaction, how much change should Sally receive?

### SOLUTION:

The correct response is:

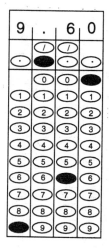

Let $x$ = the cost of 3 tangerines.

We can set up the following proportion:

$$\frac{12 \text{ tangerines}}{\$1.60} = \frac{3 \text{ tangerines}}{x}$$

$$12x = 3\,(1.60)$$

$$12x = 4.80$$

$$x = \frac{4.80}{12}$$

$$x = .40$$

Sally pays $0.40 for 3 tangerines.

If she pays with a $10.00 bill, her change is $10.00 − $0.40 = $9.60.

## • PROBLEM 3-12

A rectangular desk that measures 32 inches by 77 inches is to be completely covered by pieces of paper. Each piece of paper measures 8 inches by 11 inches. What is the least number of pieces of paper that will be required to cover the desk?

## SOLUTION:

The correct response is:

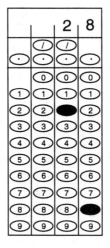

Rectangular desk measures 32 × 77

Piece of paper measures 8 × 11

It will take exactly 4 pieces of paper to cover the side of the desk that measures 32 inches.

32 ÷ 8 = 4

It will take exactly 7 pieces of paper to cover the side of the desk that measures 77 inches.

77 ÷ 11 = 7

To cover the desk entirely, we need at least (7) (4) = 28 pieces of paper.

## • PROBLEM 3–13

Five times the smaller of two whole numbers is less than one-fourth the larger number. If the value of the larger number is 84, what is the *largest* possible value of the smaller?

### SOLUTION:

The correct response is:

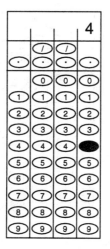

Let $x$ = the smaller number

$y$ = the larger number = 84

$$5x < \frac{1}{4} y$$

$$5x < \frac{1}{4} (84)$$

$$5x < 21$$

$$x < \frac{21}{5}$$

$$x < 4\frac{1}{5}$$

Hence, the largest possible integer value of the smaller number is 4.

## • PROBLEM 3–14

The sum of the squares of two consecutive integers is 41. What is the sum of their cubes?

### SOLUTION:

The correct response is:

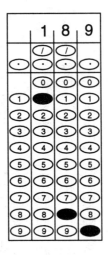

Let $x$ = the first integer

$x + 1$ = the next consecutive integer

$x^2$ = the square of the first integer

$(x + 1)^2 = x^2 + 2x + 1$ = the square of the next consecutive integer

$(x^2) + (x + 1)^2 = 41$

$(x^2) + x^2 + 2x + 1 = 41$

$2x^2 + 2x + 1 = 41$

$2x^2 + 2x - 40 = 0$

Dividing both sides of the equation by 2, we get

$x^2 + x - 20 = 0$

$(x - 4)(x + 5) = 0$

$x = 4$

$x = -5$

If $x = 4$, then $x + 1 = 5$.

If $x = -5$, then $x + 1 = -4$.

When $x = 4$, $x + 1 = 5$, the sum of their cubes is

$4^3 + 5^3 = 64 + 125 = 189.$

When $x = -5$, $x + 1 = -4$, the sum of their cubes is

$(-5)^3 + (-4)^3 = -189.$

### • PROBLEM 3-15

A class of 24 students contains 16 males. What is the ratio of females to males?

## SOLUTION:

The correct response is:

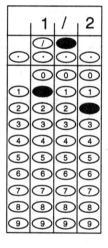

The number of females in the class is equal to the total number of students (24) minus the total number of males (16).

Total number of females = $24 - 16 = 8$.

The ratio of females to males is

$$\frac{8}{16} = \frac{1}{2}.$$

### • PROBLEM 3-16

At an office supply store, customers are given a discount if they pay in cash. If a customer is given a discount of $9.66 on a total order of $276, what is the percent of the discount?

### SOLUTION:

The correct response is:

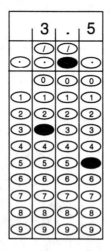

Discount = $9.66

Total order = $276

The percent of the discount is given by

$$\frac{9.66}{276} = .035.$$

To convert .035 to a percent, we move the decimal over two places to the right and attach the % sign.

Therefore, .035 = 3.5%.

## • PROBLEM 3-17

Solve for *x*.

$$x + 2y = 8$$

$$3x + 4y = 20$$

### SOLUTION:

The correct response is:

$$x + 2y = 8 \qquad\qquad (1)$$

$$3x + 4y = 20 \qquad\qquad (2)$$

We can solve for *x* in terms of *y*.

$$x + 2y = 8$$

$$x = 8 - 2y$$

We can substitute this value for *x* in equation (2).

$$3x + 4y = 20$$

$$3(8 - 2y) + 4y = 20$$

$$24 - 6y + 4y = 20$$

$$-6y + 4y = -4$$

$$-2y = -4$$

$$y = 2$$

Since $y = 2$, we can substitute this value for $y$ in equation (1) to find the value of $x$.

$$x + 2y = 8 \ (2)$$

$$x + 2(2) = 8$$

$$x + 4 = 8$$

$$x = 4$$

## • PROBLEM 3-18

Find a prime number less than 40 which is of the form $5k + 1$.

## SOLUTION:

The correct response is:

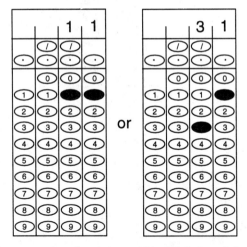

To find a prime number less than 40 which is of the form $5k + 1$, we consider the positive integers which are of the form $5k + 1$ and are less than 40.

For $\quad k = 0 \qquad 5k + 1 = 5 \ (0) + 1 = 1$

$k = 1 \qquad 5 \ (1) + 1 = 6$

$k = 2 \qquad 5 \ (2) + 1 = 11$ (prime)

$k = 3 \qquad 5 \ (3) + 1 = 16$

$k = 4 \qquad 5 \ (4) + 1 = 21$

$k = 5 \qquad 5 \ (5) + 1 = 26$

$$k = 6 \qquad 5\,(6) + 1 = 31 \text{ (prime)}$$
$$k = 7 \qquad 5\,(7) + 1 = 36$$

The prime numbers in this set are 11 and 31.

## • PROBLEM 3–19

Find the value of the expression.

$$\frac{7}{10} \times \frac{4}{21} \times \frac{25}{36} =$$

## SOLUTION:

The correct response is:

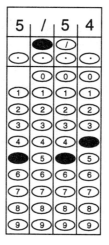

$$\frac{7}{10} \times \frac{4}{21} \times \frac{25}{36} = \frac{1}{10} \times \frac{4}{3} \times \frac{25}{36}$$

$$= \frac{1}{10} \times \frac{1}{3} \times \frac{25}{9}$$

$$= \frac{1}{2} \times \frac{1}{3} \times \frac{5}{9}$$

$$= \frac{1}{6} \times \frac{5}{9}$$

$$= \frac{5}{54}$$

## • PROBLEM 3-20

Find the solution for $x$ in the pair of equations.

$$x + y = 7$$
$$x = y - 3$$

## SOLUTION:

The correct response is:

$$x + y = 7 \tag{1}$$
$$x = y - 3 \tag{2}$$

We can substitute the value for $x$ in equation (2) into equation (1).

$$x + y = 7 \tag{1}$$
$$(y - 3) + y = 7$$
$$2y - 3 = 7$$
$$2y = 10$$
$$y = 5$$

Substituting $y = 5$ into equation (2), we get

$$x = y - 3 \tag{2}$$
$$x = 5 - 3$$
$$x = 2$$

### • PROBLEM 3-21

Simplify

$$\frac{\frac{1}{2} + \frac{1}{3}}{\frac{1}{6}}.$$

## SOLUTION:

The correct response is:

$$\frac{\frac{1}{2} + \frac{1}{3}}{\frac{1}{6}}$$

We must first simplify the numerator.

$$\frac{1}{2}\left(\frac{3}{3}\right) + \frac{1}{3}\left(\frac{2}{2}\right) = \frac{3}{6} + \frac{2}{6} = \frac{5}{6}$$

We now have

$$\frac{\frac{5}{6}}{\frac{1}{6}}.$$

$$\frac{5}{6} \div \frac{1}{6} = \frac{5}{6} \times \frac{6}{1} = 5$$

## • PROBLEM 3-22

For the triangle pictured below, the degree measures of the three angles are $x$, $3x$, and $3x + 5$. Find $x$.

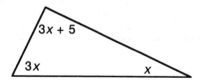

## SOLUTION:

The correct response is:

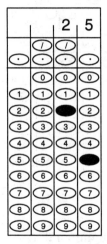

The sum of the angles of a triangle must add up to $180°$.

$$x + 3x + (3x + 5) = 180°$$

$$7x + 5 = 180°$$

$$7x = 175°$$

$$x = \frac{175°}{7} = 25°$$

## • PROBLEM 3–23

In an apartment building there are 9 apartments having terraces for every 16 apartments. If the apartment building has a total of 144 apartments, how many apartments have terraces?

### SOLUTION:

The correct response is:

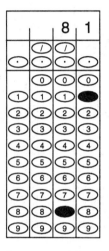

We can set up the following proportion:

$$\frac{9 \text{ terrace apartments}}{16 \text{ apartments}} = \frac{x \text{ terrace apartments}}{144 \text{ apartments}}$$

Cross multiplying we have

$$9\,(144) = 16x$$

$$1,296 = 16x$$

$$\frac{1,296}{16} = x$$

$$81 = x$$

## • PROBLEM 3-24

Solve the equation

$$2x^2 - 5x + 3 = 0.$$

## SOLUTION:

The correct response is:

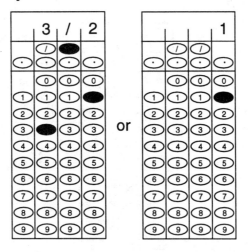

$$2x^2 - 5x + 3 = 0$$

We factor the left side of the equation, set each of these factors to zero, and solve for $x$.

$$(2x - 3)(x - 1) = 0$$

$$2x - 3 = 0 \qquad\qquad x - 1 = 0$$

$$2x = 3 \qquad\qquad x = 1$$

$$x = \frac{3}{2}$$

The solutions are $3/2$ and $1$.

## • PROBLEM 3-25

Solve the proportion

$$\frac{x+1}{4} = \frac{15}{12}.$$

### SOLUTION:

The correct response is:

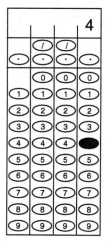

$$\frac{x+1}{4} = \frac{15}{12}$$

To solve the proportion, we cross multiply.

$$(x + 1)\,12 = 4\,(15)$$

$$12x + 12 = 60$$

$$12x = 48$$

$$x = 4$$

### • PROBLEM 3-26

What is the smallest even integer $n$ for which $(.5)^n$ is less than .01?

## SOLUTION:

The correct response is:

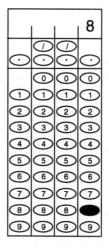

To find the smallest even integer $n$ for which $(.5)^n$ is less than .01, we start by looking at the positive even powers of .5.

When $n = 0$   $(.5)^0 = 1$

$n = 2$   $(.5)^2 = .25$

$n = 4$   $(.5)^4 = .0625$     These values are all $> .01$

$n = 6$   $(.5)^6 = .015625$

$n = 8$   $(.5)^8 = .00390625$

The answer therefore is $n = 8$.

## • PROBLEM 3-27

In the correctly worked multiplication problem below, *M*, *N*, and *P* are nonzero digits.

$$\begin{array}{r} M \\ \times\,8 \\ \hline NP \end{array}$$

If the value of *N* is 3, then what is the value of *M*?

## SOLUTION:

The correct response is:

We are given that *M*, *N*, and *P* are nonzero numbers and

$$\begin{array}{r} M \\ \times\,8 \\ \hline NP \end{array}$$

When *N* = 3

$$\begin{array}{r} M \\ \times\,8 \\ \hline 3P \end{array}$$

Thus, *M* must be a number which gives thirty-something when multiplied by 8. Trial and error is the best method here.

Try *M* = 3. This gives 3 × 8 = 24.

Try *M* = 4. This gives 4 × 8 = 32.

Try $M = 5$. This gives $5 \times 8 = 40$.

Thus, the only possible answer is $M = 4$.

### • PROBLEM 3-28

A chef bought a jar that holds 76.25 gallons of liquid. How many quart contains are required to fill the jar? (1 gallon = 4 quarts)

## SOLUTION:

The correct response is:

There are 4 quarts in a gallon. To find the number of quarts in the jar, we must multiply the number of quarts in a gallon (4) by the number of gallons in the jar (76.25).

$$4 \times (76.25) = 305 \text{ quart containers}$$

## • PROBLEM 3-29

The following shows the bank account balance of Mr. Jones on various dates.

| | |
|---|---|
| December 1 | + $75.50 |
| December 8 | − $53.36 |
| December 15 | + $62.28 |

What, to the nearest dollar, is the arithmetic mean of the account balances?

### SOLUTION:

The correct response is:

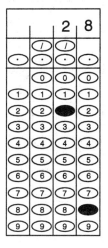

To find the arithmetic mean, we must add the bank account balances and then divide by the number of dates.

$$\text{mean} = \frac{(\$75.50) + (-\$53.36) + (\$62.28)}{3}$$

$$\text{mean} = \frac{\$84.42}{3}$$

$$\text{mean} = \$28.14$$

To the nearest dollar, $28.14 rounds off to $28.

### • PROBLEM 3–30

If

$$m^{15} = \frac{48}{y} \text{ and } m^{13} = \frac{2}{6y} \text{ and } m > 0,$$

what is the value of $m$?

## SOLUTION:

The correct response is:

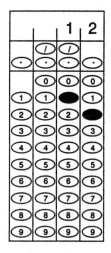

$$m^{15} = \frac{48}{y}$$

$$m^{13} = \frac{2}{6y}$$

$m^{15}$ and $m^{13}$ are powers of the same base.

$$\frac{m^{15}}{m^{13}} = m^2 = \frac{\frac{48}{y}}{\frac{2}{6y}} = \frac{48}{y} \div \frac{2}{6y} = \frac{48}{y} \times \frac{6y}{2}$$

$$m^2 = 144$$

$$m = \sqrt{144} = 12$$

### • PROBLEM 3-31

When 8 consecutive integers are multiplied, their product is 0. What is their maximum sum?

## SOLUTION:

The correct response is:

If the product of 8 integers is zero, we know that at least one of the integers must be zero.

Since the integers are *consecutive*, no more than one of the integers can equal zero.

Here are a few examples of sets of integers which meet the above requirements:

$$-7, -6, -5, -4, -3, -2, -1, 0$$

and     $-3, -2, -1, 0, 1, 2, 3, 4$

and     $0, 1, 2, 3, 4, 5, 6, 7$

The problem asks for maximum sum.

The more negative numbers we include, the smaller our sum will be.

Thus, the maximum sum equals

$$0 + 1 + 2 + 3 + 4 + 5 + 6 + 7 + 8 = 28$$

## • PROBLEM 3-32

Four numbers are selected at random. They have an average (arithmetic mean) of 47. The third number selected was 12. What is the sum of the other three numbers?

### SOLUTION:

The correct response is:

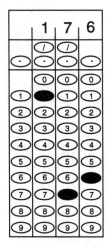

If four numbers have an average of 47, their sum would be the product of the average multiplied by the number of numbers.

47 × 4 = 188

If the third number equals 12 then the sum of the remaining three numbers must equal the total sum minus 12.

188 − 12 = 176

The sum of the other three numbers equals 176.

## • PROBLEM 3-33

$$\frac{y}{27} = \frac{3}{y}$$

What is the largest possible value of $y$ that would solve the equation above?

## SOLUTION:

The correct response is:

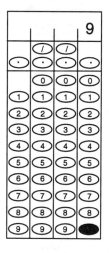

$$\frac{y}{27} = \frac{3}{y}$$

In order to solve this problem for $y$, we must cross multiply.

$y^2 = 3 (27)$

$y^2 = 81$

$y = \sqrt{81} = \pm 9$

We find that $y$ can equal $+ 9$ or $- 9$. The larger value is $y = 9$.

### • PROBLEM 3-34

Square *A* has a side of $\sqrt{6}$ and square *B* has a side of 6. How much greater is the area of square *B* than the area of square *A*?

## SOLUTION:

The correct response is:

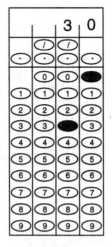

To find out how much greater the area of square *B* is than the area of square *A*, we must first find the area of each square, and then subtract the area of square *A* from the area of square *B*.

Area of a square = (side)$^2$

Area of square $A = (\sqrt{6})^2 = 6$

Area of square $B = (6)^2 = 36$

Area of Square *B* – Area of Square *A* $= 36 - 6 = 30$

## • PROBLEM 3-35

The product of 7 and $n$ is 11 more than $m$. If $n = 6$, what is the value of $m$?

## SOLUTION:

The correct response is:

We know that the product of 7 and $n$ is 11 more than $m$. We can set up the following equation:

$$7 \times n = m + 11$$

We can now substitute $n = 6$ into the equation.

$$7 \times (6) = m + 11$$

We can now simplify the equation and solve for $m$.

$$42 = m + 11$$
$$42 - 11 = m + 11 - 11$$
$$31 = m$$

## • PROBLEM 3-36

What would 20% of 3x be, if 60% of x were 15?

### SOLUTION:

The correct response is:

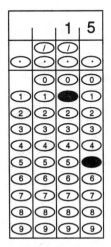

If 60% of $x$ is 15, we get

$$60\% \times x = 15$$

$$.60\, x = 15$$

$$x = \frac{15}{.60}$$

$$x = 25$$

3x is then 3 (25) = 75.

20% of 3x = 20% of 75

$$= .20 \times 75$$

$$= 15$$

## • PROBLEM 3-37

A gym teacher enters the swimming pool area at 8:30 A.M. and notices that the water in the pool is only 1.5 feet deep. State regulations require that the water be at least 5 feet deep. If the water flowing into the pool raises the water level at a rate of .25 feet per hour, how many hours will it take to fill the pool to the height required by state law?

## SOLUTION:

The correct response is:

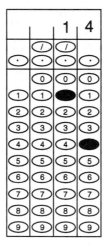

If the water is 1.5 feet deep and it must be 5 feet deep, the gym teacher must add 5 − 1.5 = 3.5 feet of water.

If the water raises the level .25 feet per hour and we want to find out how long it would take to raise the water level by 3.5 feet, we must divide the amount we need the water raised (3.5 feet) by the amount the water will be raised each hour (.25 feet).

3.5 ÷ .25 = 14

It will take 14 hrs.

• **PROBLEM 3-38**

If $a = 10$, $b = 50$, and $c = a \times b$, what is the decimal value of $\frac{1}{c}$?

## SOLUTION:

The correct response is:

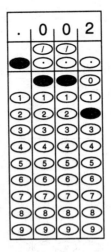

$a = 10$

$b = 50$

$c = a \times b$

$c = 10 \times 50$

$c = 500$

Therefor,e

$$\frac{1}{c} = \frac{1}{500}$$

To express $^1/_{500}$ as a decimal, we divide the numerator by the denominator

$1.000 \div 500 = .002$

Therefore, $\frac{1}{c} = .002$

### • PROBLEM 3–39

What are the number of minutes in a day, rounded to the nearest hundred?

## SOLUTION:

The correct response is:

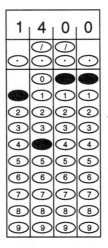

One day has 24 hours.

One hour has 60 minutes.

To find out how many minutes in a day, we multiply the number of hours in a day by 60.

24 hours/day × 60 min/hour = 1,440 min/day

Rounded to the nearest hundred, we have 1,400 minutes.

## • PROBLEM 3-40

In a school, 18 students take math and 12 take biology. If there are a total of 22 students enrolled in the school, how many students take both math and biology?

## SOLUTION:

The correct response is:

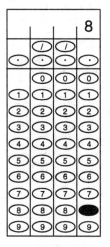

Let $x$ = number of students who take math and biology

$y$ = number taking math alone

$z$ = number taking biology alone

$$x + y = 18$$

$$x + z = 12$$

$$x + y + z = 22$$

We can substitute $(x + y) = 18$ into the last equation.

$$18 + z = 22$$

$$z = 4$$

Four equals the number of students taking biology alone.

If $z = 4$ and $x + z = 12$, we can solve for $x$.

$$x + z = 12$$

$$x + 4 = 12$$

$$x = 8$$

This is the number of students who are taking both math and biology.

### • PROBLEM 3–41

The mean (average) of the numbers 50, 60, 65, 75, $x$, and $y$ is 65. What is the mean of $x$ and $y$?

## SOLUTION:

The correct response is:

If the average of the 6 numbers is 65, we have the following equation:

$$\frac{50 + 60 + 65 + 75 + x + y}{6} = 65$$

$$250 + x + y = (65)\,6$$

$$x + y = 390 - 250$$

$$x + y = 140$$

If the sum of $x$ and $y$ is 140, then the mean of $x$ and $y$ is

$$\frac{140}{2} = 70.$$

## • PROBLEM 3–42

The ages of the students enrolled at XYZ University are given in the following table:

| Age | Number of students |
|-----|--------------------|
| 18  | 750                |
| 19  | 1,600              |
| 20  | 1,200              |
| 21  | 450                |

What percent of students are 19 and 20 years old?

## SOLUTION:

The correct response is:

$$\frac{\text{number of 19 year olds} + \text{number of 20 year olds}}{\text{total number of students}} =$$

$$= \frac{1,600 + 1,200}{750 + 1,600 + 1,200 + 450} = \frac{2,800}{4,000}$$

$$= \frac{7}{10} = .70$$

To find the percent, we move the decimal point two places to the right and add the % sign.

$$\frac{7}{10} = .70 = 70\%$$

## • PROBLEM 3-43

Find the larger side of a rectangle whose area is 24 and whose perimeter is 22.

## SOLUTION:

The correct response is:

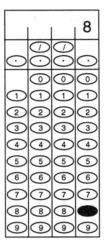

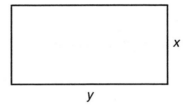

$y$

Let $x$ = smaller side of the rectangle (length)

Let $y$ = larger side of the rectangle (width)

Area = (length) (width)

$24 = xy$

Perimeter = 2 (length) + 2 (width)

$22 = 2x + 2y$

We can simplify this equation by dividing both sides by 2.

$11 = x + y$

We can now solve for $y$ in terms of $x$. We see that

$$y = (11 - x)$$

Substituting this value for $y$ in the area equation, we get

$$24 = x \, (11 - x)$$
$$24 = 11x - x^2$$

We can now set one side equal to zero.

$$x^2 - 11x - 24 = 0$$

We can now factor and set each of the factors equal to zero.

$$(x - 8) \, (x - 3) = 0$$

$$x - 8 = 0 \qquad x - 3 = 0$$
$$x = 8 \qquad\quad x = 3$$

If $x = 8$, then $y = 3$

since
$$xy = 24$$
$$8y = 24$$
$$y = 3$$

Likewise:

If $x = 3$, then $y = 8$.

The value of the larger side is therefore 8.

## • PROBLEM 3-44

The following are three students' scores on Mr. Page's Music Funda-mentals midterm. The given score is the number of correct answers out of 55 total questions.

| Liz | 48 |
| Jay | 45 |
| Carl | 25 |

What is the *average* percentage of questions correct for the three students?

## SOLUTION:

The correct response is:

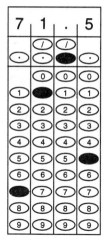

To find the average number of correct questions, we must add the scores obtained by each student and then divide by the number of students.

$$\frac{48 + 45 + 25}{3} = \frac{118}{3} = 39\frac{1}{3}$$

$39^1/_3$ is the average number of correct questions.

To find the average percentage of correct questions, we divide the average number of correct questions by the total number of questions.

$$\frac{39\frac{1}{3}}{55} = \frac{\frac{118}{3}}{55} = \frac{118}{165} = .715 = 71.5\%$$

## • PROBLEM 3-45

Reserved seat tickets to a football game are $6 more than general admission tickets. Mr. Jones finds that he can buy general admission tickets for his whole family of five for only $3 more than the cost of reserved seat tickets for himself and Mrs. Jones. How much do the general admission tickets cost?

## SOLUTION:

The correct response is:

Let $x$ = price of a general admission seat

$x + 6$ = price of a reserved seat

Therefore $5x$ = price of general admission seats for five people

$2(x + 6)$ = cost of reserved seats for Mr. and Mrs. Jones

Thus, we can say

$5x = 2(x + 6) + 3$

$5x = 2x + 12 + 3$

$5x = 2x + 15$

$3x = 15$

$x = 5$

$5.00 is the price for general admission seats.

## • PROBLEM 3–46

The sum of three numbers is 96. The ratio of the first to the second is 1 : 2, and the ratio of the second to the third is 2 : 3. What is the third number?

## SOLUTION:

The correct response is:

Let $x$ = first number

$y$ = second number

$z$ = third number

$$x + y + z = 96$$

$$\frac{x}{y} = \frac{1}{2}$$

$$\frac{y}{z} = \frac{2}{3}$$

Solving for $y$ in the third equation, we see that

$$y = \frac{2z}{3}.$$

Likewise, solving for $x$ in the second equation, we get

$$x = \frac{y}{2} = \frac{\frac{2z}{3}}{2} = \frac{z}{3}.$$

Plugging both of these terms into the first equation, we obtain

$$\frac{z}{3} + \frac{2z}{3} + z = 96$$

$$2z = 96$$

$$z = 48$$

## • PROBLEM 3-47

If $L_1 \parallel L_2$ and $L_4$ and $L_1$ are tangents to the circle, then what does angle $f$ equal?

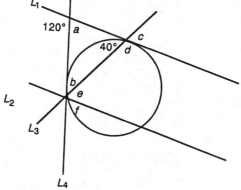

Note: Figure not drawn to scale.

## SOLUTION:

The correct response is:

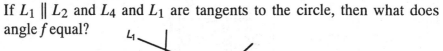

The sum of the angles of a line is 180°. Because angle $a$ and the 120°

angle form $L_1$, they must add up to 180°.

$$a + 120° = 180°$$

$$a = 60°$$

If $L_1 \parallel L_2$, then angle $a$ is equal to angle $f$. This is because corresponding angles of parallel lines are equal.

Therefore, $f = 60°$

## • PROBLEM 3-48

Last Valentine's Day 100 red roses cost $r$ dollars. This Valentine's Day, 80 of the same kind of red rose cost $.2r$ dollars. If there were no discounts based on the size of the purchase, what was the percentage decrease in the cost of a red rose?

### SOLUTION:

The correct response is:

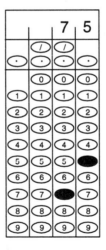

80 red roses cost $.2r$ dollars.

$$80 = .2r$$

$$\frac{80}{.2} = r$$

$$400 = r$$

This means that this year $r$ dollars will buy 400 roses. To find the percent change, we must first compute how much 100 red roses would cost at this year's price.

$$\frac{r}{400} = \frac{x}{100}$$

$$100r = 400x$$

$$\frac{100r}{400} = x$$

$$\frac{r}{4} = x$$

$$.25r = \frac{r}{4} = x$$

Since 100 cost $r$ dollars last year, but 100 roses cost $.25r$ this year, the percentage decrease in value of a rose would be the difference in value between the cost of 100 roses last year and 100 roses this year.

$$1.00r - .25r = .75r$$

This value represents a decrease of 75%.

## • PROBLEM 3-49

In the figure shown below, if the average (arithmetic mean) of $\overline{FG}$ and $\overline{GH}$ is 12.2, what is the distance from point $G$ to point $H$?

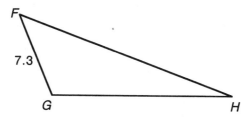

Note: Figure not drawn to scale.

## SOLUTION:

The correct response is:

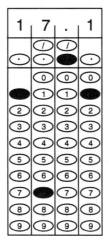

$$\text{Average} = \frac{\overline{FG} + \overline{GH}}{2} = 12.2$$

$\overline{FG} + \overline{GH} = 24.4$

Since $\overline{FG} = 7.3$

$7.3 + \overline{GH} = 24.4$

Therefore, solving for $\overline{GH}$, we get

$\overline{GH} = 24.4 - 7.3 = 17.1$.

### • PROBLEM 3–50

The list price of a new car was $10,000. During a sale, the car was sold for $7,500. By what percent was the original price of the car reduced?

## SOLUTION:

The correct response is:

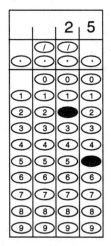

Let $x$ = percent of decrease

$$\frac{\text{amount of decrease}}{\text{original amount}} = x\%$$

$$\frac{10,000 - 7,500}{10,000} = \frac{x}{100}$$

$$\frac{2,500}{10,000} = \frac{x}{100}$$

$$100x = 2,500$$

$$x = 25$$

## • PROBLEM 3-51

Two consecutive even integers have a sum of 26. What is the result when they are multiplied?

## SOLUTION:

The correct response is:

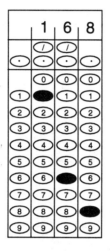

Let $x$ = first even integer

$x + 2$ = second even integer

$x + (x + 2) = 26$

$2x + 2 = 26$

$2x = 24$

$x = 12$

$x + 2 = 14$

$(x)(x + 2) = (12)(14) = 168$

## • PROBLEM 3-52

On Sunday there are 15 restaurants open in the city. Each restaurant serves an average (arithmetic mean) of 1,200 customers. On the following day, 7 of the restaurants close for a holiday, but the same number of people use the restaurants. What is the increase in the average number of customers at each restaurant?

## SOLUTION:

The correct response is:

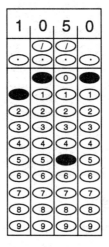

If the 15 restaurants served an average of 1,200 customers, then the 15 restaurants served a total of

15 × 1,200 or 18,000 customers.

If 7 of 15 restraurants close, then 8 would remain open.

If 18,000 customers were served by 8 restaurants, then each served an average of

$$\frac{18,000}{8} \text{ or 2,250 customers.}$$

If on Sunday each restaurant averaged 1,200 customers and on Monday each restaurant averaged 2,250 customers, the increase in the average number of customers at each restaurant is

2,250 – 1,200 = 1,050 customers.

## • PROBLEM 3-53

An ancient map indicates that treasure is buried 300 feet below the bottom of a mountain 1,500 feet high, the peak of which is 2,750 feet above the bottom of a cliff located close to the mountain. If the map is accurate, how many feet above the bottom of the cliff is the treasure located?

## SOLUTION:

The correct response is:

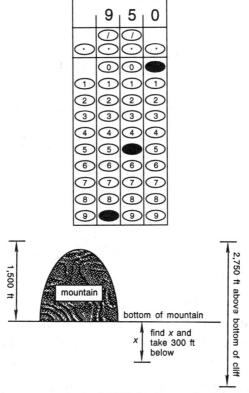

The peak of the mountain is 2,750 ft above the bottom of the cliff.

The mountain is 1,500 ft high. Therefore, the height of the bottom of the mountain relative to the bottom of the cliff is

$$2,750 - 1,500 = 1,250.$$

The treasure was buried 300 feet below the bottom of the mountain.

$$1,250 - 300 = 950$$

This is the location of the treasure relative to the bottom of the cliff.

## • PROBLEM 3-54

Define ¥ by the following equations:

$m \text{ ¥ } n = 2m + n$, where $n > 0$.

$m \text{ ¥ } n = m - 2n$, where $n \leq 0$.

What is the value of the following expression:

$6 \text{ ¥ } - 4$?

## SOLUTION:

The correct response is:

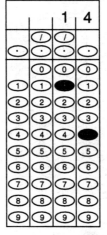

$m \text{ ¥ } n = 2m + n$, where $n > 0$

$m \text{ ¥ } n = m - 2n$, where $n \leq 0$

We must use the fact that the symbol ¥ changes according to the value of $n$. In the expression $6 \text{ ¥ } -4$, the number to the right of ¥ is $-4$. This is less than zero. We must therefore use the second equation.

The second equation says that we are to take the number to the left of ¥ (in this case 6) and from that number subtract twice the number to the right of ¥ (in this case $-4$).

$m - 2n$

$6 - 2(-4)$

$6 - (-8)$

$6 + 8 = 14$

### • PROBLEM 3-55

If *m* is an integer and the sum of *m* and the next integer larger than *m* is greater than 10, what is the smallest possible value of *m*?

## SOLUTION:

The correct response is:

Let $m$ = an integer

$m + 1$ = the next larger integer

$$m + (m + 1) > 10$$

$$2m + 1 > 10$$

$$2m > 10 - 1$$

$$2m > 9$$

$$m > \frac{9}{2}$$

$$m > 4.5$$

The smallest integer greater than 4.5 is 5.

## • PROBLEM 3-56

If the sum of the digits of a two-digit positive whole number is 8 and the tens digit is three times the units digit, then what is the two-digit number?

### SOLUTION:

The correct response is:

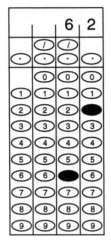

Let $u$ = the units digit

$t$ = the tens digit

$u + t = 8$

$t = 3u$

Substituting $t = 3u$ for $t$ in $u + t = 8$, we get

$u + t = 8$

$u + 3u = 8$

$4u = 8$

$u = 2$

The units digit is 2.

Substituting this value for $u$ into either of the equations above, we get

$u + t = 8$      $t = 3u$

$2 + t = 8$      $t = 3 \times 2$

$t = 6 \quad t = 6$

The tens digit is 6.

The number therefore is 62.

## • PROBLEM 3-57

In the diagram below, $L1 \parallel L2$ and $L3 \parallel L4$. If $a + d + b = 8e$, what is the measure of angle $e$?

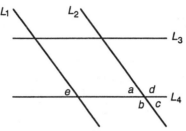

## SOLUTION:

The correct response is:

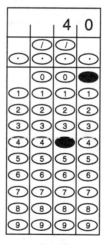

$\angle c = \angle a$

vertical angles are equal.

$\angle e = \angle a$

corresponding angles of parallel lines are equal.

$\angle c = \angle e$

since angles equal to the same angle are equal to each other.

$$\angle a + \angle b + \angle c + \angle d = 360$$

since $\angle c + \angle d = 180$

and $\angle a + \angle b = 180$.

We are given that

$$\angle a + \angle d + \angle b = 8(\angle e).$$

Then substituting we get

$$8(\angle e) + \angle c = 360°$$

$$8(\angle e) + \angle e = 360°$$

$$9(\angle e) = 360°$$

$$\angle e = \frac{360°}{9} = 40°$$

## • PROBLEM 3–58

Find the length of a side of an equilateral triangle whose area is $4\sqrt{3}$.

## SOLUTION:

The correct response is:

The area of an equilateral triangle is

$\frac{1}{2}bh.$

Using the Pythagorean Theorem we have

$$b^2 = \left(\frac{b}{2}\right)^2 + h^2$$

$$b^2 = \frac{b^2}{4} + h^2$$

$$\frac{3b^2}{4} = h^2$$

$$\frac{\sqrt{3}b}{2} = h$$

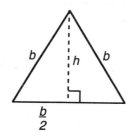

Substituting

$$h = \frac{\sqrt{3}b}{2}$$

into  $\quad \frac{1}{2}bh = 4\sqrt{3}$

$$\frac{1}{2}\frac{\sqrt{3}b^2}{2} = 4\sqrt{3}$$

$$b^2 = 16$$

$$b = 4$$

## • PROBLEM 3-59

Solve

$$\frac{3}{x-1} + \frac{1}{x-2} = \frac{5}{(x-1)(x-2)}.$$

## SOLUTION:

The correct response is:

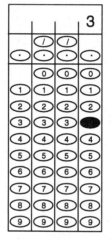

$$\frac{3}{x-1} + \frac{1}{x-2} = \frac{5}{(x-1)(x-2)}$$

$$(x-1)(x-2)\frac{3}{x-1} + (x-1)(x-2)\frac{1}{x-2} = \frac{5}{(x-1)(x-2)}(x-1)(x-2)$$

This simplifies to

$$3(x-2) + (x-1) = 5$$

$$3x - 6 + x - 1 = 5$$

$$4x - 7 = 5$$

$$4x = 12$$

$$x = 3$$

## • PROBLEM 3-60

In an isosceles triangle, the length of each of the congruent sides is 10 and the length of the base is 12. Find the length of the altitude drawn to the base.

## SOLUTION:

The correct response is:

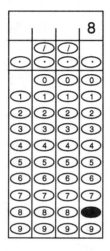

In order to find the altitude $A$, we use the Pythagorean Theorem.

The altitude divides the base in half.

$$6^2 + A^2 = 10^2$$

$$36 + A^2 = 100$$

$$A^2 = 64$$

$$A = 8$$

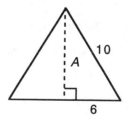

## • PROBLEM 3-61

$\triangle PQR$ is a scalene triangle. The measure of $\angle P$ is eight more than twice the measure of $\angle R$. The measure of $\angle Q$ is two less than three times the measure of $\angle R$. Determine the measure of $\angle Q$.

### SOLUTION:

The correct response is:

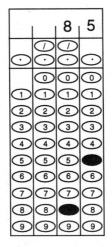

In a scalene triangle the three sides are not equal.

Let $x$ = measure of $\angle R$

$2x + 8$ = measure of $\angle P$

$3x - 2$ = measure of $\angle Q$

Since the sum of the angles in a triangle = 180°, then

$$\angle P + \angle Q + \angle R = 180°$$
$$(2x + 8) + (3x - 2) + x = 180°$$
$$6x + 6 = 180$$
$$6x = 174$$
$$x = 29$$

Therefore,

$$\angle Q = 3x - 2$$
$$= 3\,(29) - 2$$
$$= 87 - 2$$
$$= 85°$$

## • PROBLEM 3-62

A mother is now 24 years older than her daughter. In 4 years, the mother will be 3 times as old as the daughter. What is the present age of the daughter?

## SOLUTION:

The correct response is:

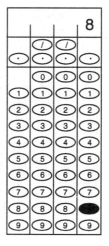

Let $x$ = daughter's age now

$x + 4$ = daughter's age in 4 years

$x + 24$ = mother's age now

$(x + 24) + 4$ = mother's age in 4 years

$(x + 24) + 4 = 3 (x + 4)$

$x + 28 = 3x + 12$

$28 = 2x + 12$

$16 = 2x$

$8 = x$

The present age of the daughter is 8.

## • PROBLEM 3-63

John is four times as old as Harry. In six years John will be twice as old as Harry. What is Harry's age now?

## SOLUTION:

The correct response is:

Let $x$ = Harry's age now

$4x$ = John's age now

$x + 6$ = Harry's age in 6 years

$4x + 6$ = John's age in 6 years

$$4x + 6 = 2(x + 6)$$

$$4x + 6 = 2x + 12$$

$$2x = 6$$

$$x = 3$$

Harry is now 3 years old.

## • PROBLEM 3-64

What is the area of the shaded region?

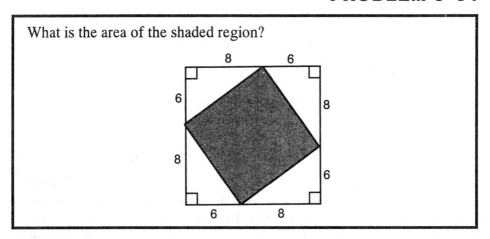

## SOLUTION:

The correct response is:

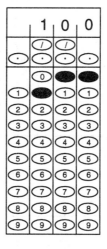

The area of the shaded region is the total area of the square minus the area of the four right triangles.

The area of the square = $(\text{side})^2$.

The side of the square is $(8 + 6) = 14$.

The area of the square is $14^2 = 196$.

The areas of the triangles can be determined using the equation

$$\frac{1}{2}bh.$$

The area of the triangles is

$\frac{1}{2}(8)(6) = 24.$

Since there are four triangles, the total area of the triangles is

$4(24) = 96.$

The area of the shaded region is therefore

$196 - 96 = 100.$

## • PROBLEM 3-65

In the diagram shown, $ABC$ is an isosceles triangle. Sides $\overline{AC}$ and $\overline{BC}$ are extended through $C$ to $E$ and $D$ to form triangle $CDE$. What is the sum of the measures of angles $D$ and $E$?

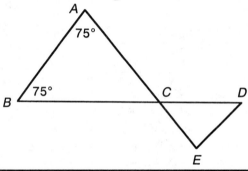

## SOLUTION:

The correct response is:

The sum of the angles in a triangle is 180°.

$$\angle CAB + \angle ABC + \angle BCA = 180°$$

$$75° + 75° + \angle BCA = 180°$$

$$150° + \angle BCA = 180$$

$$\angle BCA = 30°$$

Since $\angle ACB$ is a vertical angle to $\angle DCE$, then $\angle ACB = \angle DCE = 30$.

In triangle $CDE$ we have

$$\angle D + \angle E + \angle DCE = 180°$$

$$\angle D + \angle E + 30° = 180°$$

$$\angle D + \angle E = 150°$$

## • PROBLEM 3–66

| Number of Muffins | Total Price |
|-------------------|-------------|
| 1 | $0.55 |
| Box of 4 | $2.10 |
| Box of 8 | $4.00 |

According to the information in the table above, what would be the *least* amount of money needed to purchase exactly 19 muffins? (Disregard the dollar sign when gridding your answer.)

## SOLUTION:

The correct response is:

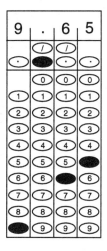

In order to make good use of the discount we want to first buy boxes of eight, then boxes of four, and then buy single muffins. This minimizes the price.

Two boxes of eight muffins gives us a total of

$2 \times 8 = 16$ muffins, plus $3 \times$ single muffins

would give us the 19 muffins.

$$2 (\$4.00) + 3 (\$0.55) = \$8.00 + \$1.65$$

$$= \$9.65$$

### • PROBLEM 3–67

Several people rented a van for $30, sharing the cost equally. If there had been one more person in the group, it would have cost each $1 less. How many people were there in the group originally?

## SOLUTION:

The correct response is:

Let $x$ = number of people renting the van

$y$ = cost to each person

$xy = 30$

By adding one more person to the van, the cost per person goes down by $1.

This means that

$x + 1$ = new number of people renting the van for $30

$y - 1$ = new cost per person

Now

$(x + 1)(y - 1) = 30.$

Substituting $xy = 30$ into this equation gives

$(x + 1)(y - 1) = xy$

$xy - x + y - 1 = xy$

$-x + y - 1 = 0$

$y = x + 1$

Since $xy = 30$, we can substitute $y = x + 1$ into this equation.

$x(x + 1) = 30$

$x^2 + x = 30$

$x^2 + x - 30 = 0$

$(x + 6)(x - 5) = 0$

$x + 6 = 0 \qquad\qquad x - 5 = 0$

$x = -6 \qquad\qquad x = 5$

Since the number of people in the van cannot be a negative number, the answer is 5.

### • PROBLEM 3-68

Find the area of the shaded triangles.

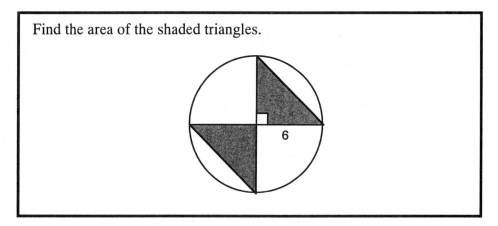

## SOLUTION:

The correct response is:

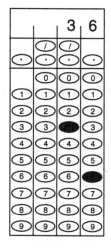

We are given that the radius of the circle is 6. The area of the shaded region is the area of the two triangles.

The area = $\frac{1}{2}bh$.

The base and height of the triangle equal the radius.

$\frac{1}{2}bh = \frac{1}{2}(6)(6) = 18$

The area of two triangles is

$2 \times 18 = 36$.

## • PROBLEM 3-69

Given the rhombus *RHOM,* find the length of the diagonal $\overline{RO}$.

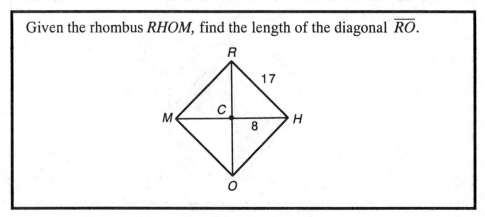

## SOLUTION:

The correct response is:

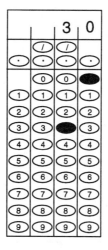

The diagonals of a rhombus meet at right angles.

Since $\overline{CH} = 8$ and $\overline{RH} = 17$, we can solve for $\overline{RC}$ using the Pythagorean Theorem as follows:

$$(\overline{CH})^2 + (\overline{RC})^2 = (\overline{RH})^2$$
$$8^2 + (\overline{RC})^2 = 17^2$$
$$64 + (\overline{RC})^2 = 289$$
$$(\overline{RC})^2 = 289 - 64$$
$$(\overline{RC})^2 = 225$$
$$\overline{RC} = 15$$

Therefore,

$$\overline{RO} = \overline{RC} + \overline{CO}$$
$$\overline{RO} = 15 + 15 = 30$$

## • PROBLEM 3-70

$\triangle MNO$ is isosceles. If the vertex angle, $\angle N$, has a measure of 96°, find the measure of $\angle M$.

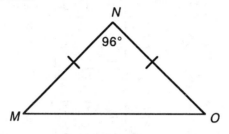

### SOLUTION:

The correct response is:

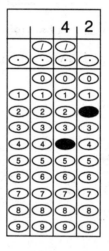

In an isosceles triangle two sides are equal.

$$\overline{MN} = \overline{NO} \text{ and } \angle M = \angle O$$

Since the sum of the angles of a triangle equals 180, we have

$$\angle N + \angle M + \angle O = 180°$$

$$96° + 2(\angle M) = 180°$$

$$2(\angle M) = 180° - 96°$$

$$2(\angle M) = 84°$$

$$\angle M = 42°$$

## • PROBLEM 3-71

The sum of two integers is 31, and the difference between the two integers is 7. What is the larger of the two integers?

**SOLUTION:**

The correct response is:

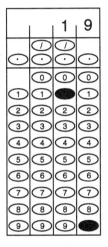

Let $x$ = first integer

$y$ = second integer

$$x + y = 31 \tag{1}$$

$$x - y = 7 \tag{2}$$

Wc can solve equation (1) for $x$.

$$x + y = 31$$

$$x = 31 - y$$

We can now substitute this value for $x$ in equation (2).

$$x - y = 7$$

$$(31 - y) - y = 7$$

$$31 - 2y = 7$$

$$24 = 2y$$

$$12 = y$$

We can now substitute this value into either equation (1) or equation

(2) to give us the value of $x$.

$$x + y = 31$$

$$x + 12 = 31$$

$$x = 31 - 12 = 19$$

Since $12 < 19$, the larger of the two integers is 19.

## • PROBLEM 3-72

In the figure below, if $\angle e = 135°$, what is the value of $\angle f + \angle g$?

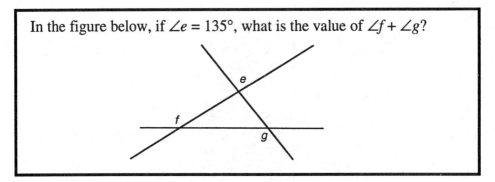

## SOLUTION:

The correct response is:

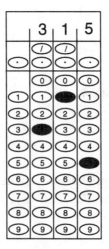

$\angle e = 135°$

$\angle e = \angle i$

because vertical angles are equal.

$\angle h + \angle f = 180$

$\angle j + \angle g = 180$

Supplementary angles add up to 180°.

$\angle h = 180 - \angle f$

$\angle j = 180 - \angle g$

$\angle i + \angle j + \angle h = 180$

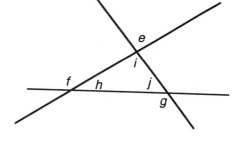

The sum of the angles of a triangle is 180°.

Substituting:

$135 + (180 - \angle g) + (180 - \angle f) = 180°$

$495 - \angle g - \angle f = 180°$

$495 - 180 - \angle g - \angle f = 0$

$315 = \angle g + \angle f$

## • PROBLEM 3–73

If $1/4$ of an even positive number and $5/6$ of the next larger even number have a sum of 32, what is the average of the two numbers?

## SOLUTION:

The correct response is:

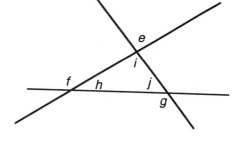

Let $x$ = first even positive number

$x + 2$ = second even positive number

$$\frac{1}{4}x + \frac{5}{6}(x + 2) = 32$$

We must now solve for $x$.

$$\frac{1}{4}x + \frac{5x + 10}{6} = 32$$

$$\frac{3}{3}\left(\frac{1}{4}x\right) + \frac{2}{2}\left(\frac{5x + 10}{6}\right) = 32$$

$$\frac{3x}{12} + \frac{10x + 20}{12} = 32$$

$$\frac{3x + 10x + 20}{12} = 32$$

$$13x + 20 = (32)\,12$$

$$13x + 20 = 384$$

$$13x = 384 - 20$$

$$13x = 364$$

$$x = \frac{364}{13}$$

$$x = 28$$

The numbers are 28 and 30.

The average of the two numbers is

$$\frac{28 + 30}{2} = 29.$$

### • PROBLEM 3-74

A crop duster reading a map learns that on the map she must treat a circular area of $9\pi$ sq. in. If 30 miles on the ground is represented by 1 inch on the map, the area in square miles that the crop duster must treat is _____ $\pi$?

## SOLUTION:

The correct response is:

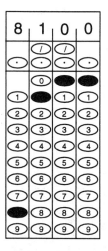

The area on the map that the crop duster must treat is $9\pi$ sq in. We can compute the radius of the circle as follows:

Area of circle $= \pi r^2$

$9\pi = \pi r^2$

$9 = r^2$

$3 = r$

The radius on the map is 3 inches. Since 1 inch is equal to 30 miles, we can set up a proportion to determine the radius of the circle on the ground.

$$\frac{30}{1} = \frac{x}{3}$$

$$x = 3\,(30)$$

$$x = 90$$

Thus, the radius of the circle on the ground is 90. Using this information we can now compute the area of the circle on the ground.

Area of circle $= \pi \rho^2$

Area $= \pi(90)^2$

Area $= 8{,}100\pi$ sq. miles

### • PROBLEM 3-75

A soccer team has played 22 games of which it has won 13 and lost 9. If the team wins its remaining $z$ games and finishes the season with a 70% winning percentage, what is the value of $z$?

## SOLUTION:

The correct response is:

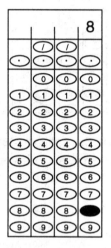

The number of games won divided by the total number of games played is equal to the winning percentage.

$$\frac{\text{number of games won}}{\text{total number of games played}} = \text{winning percentage}$$

$$\frac{13+z}{22+z} = 70\%$$

$$\frac{13+z}{22+z} = .70$$

$$\frac{13+z}{22+z} = \frac{70}{100}$$

$$100\,(13+z) = 70\,(22+z)$$

$$1{,}300 + 100z = 1{,}540 + 70z$$

$$30z = 240$$

$$z = \frac{240}{30}$$

$z = 8$

The team must win the next 8 games.

## • PROBLEM 3–76

What would be the value of $y$ if $\sqrt{81} = 3^y$?

## SOLUTION:

The correct response is:

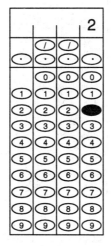

$\sqrt{81} = 3^y$ $\qquad\qquad$ $\sqrt{81} = 9$

$\quad 9 = 3^y$

We must write 9 with a base of 3 to solve for $y$.

$\quad 3^2 = 3^y$

$\quad 2 = y$

### • PROBLEM 3-77

A student received a grade of A on 6 out of every 36 tests she took during her junior year. In fractional form, what was the ratio of non-A to A grades that the student received during her junior year?

## SOLUTION:

The correct response is:

We know that the student received a grade of A on 6 tests out of a total of 36 tests. The student received a non-A grade on 36 − 6 tests = 30 tests.

Therefore the ratio

$$\frac{\text{non-A grades}}{\text{A grades}} = \frac{30}{6} = \frac{5}{1}$$

## • PROBLEM 3-78

Eleven minutes is what fractional part of 11 hours?

## SOLUTION:

The correct response is:

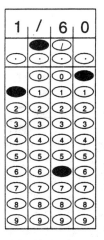

$$11 \text{ hours} \times 60 \text{ min/hr} = (11)(60) \text{ min}$$

$$= 660 \text{ min}$$

Hence, eleven minutes is

$\dfrac{11}{660}$ of 11 hours.

So eleven minutes is $1/60$ of 11 hours.

## • PROBLEM 3-79

The PTA of a certain school has 150 members. The PTA wishes to form a number of committees on which exactly 10 people will serve. If no one may serve on more than two committees, what is the maximum number of committees that may be formed?

### SOLUTION:

The correct response is:

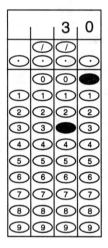

If the PTA has 150 members which will be divided up into committees of 10 people, then the number of committees that could be formed would be

$$\frac{150}{10} = 15 \text{ committees}$$

Since no one person may serve on more than two committees, then each of the 15 committees could create two committees. There would then be $15 \times 2 = 30$ possible committees.

## • PROBLEM 3-80

To induce comraderie, a teacher has decided that each member of his English class will shake the hand of every other student of the English class at the beginning of each class. If there are 12 students in attendance, how many handshakes will there be?

## SOLUTION:

The correct response is:

One way to visualize the problem is that one person shakes everyone's hand and then leaves the room. Then a second person shakes the hand of everyone in the room, and then that person leaves the room. We continue to do this until there is one person left in the room.

The first person will shake 11 hands. The second person will shake 10 hands. With 10 students in the room, the third student will shake 9 hands. We can see that a pattern is developing. Each time we increase the number of students doing the handshaking by one, we also decrease the number of handshakes by one.

| # handshakes | 11 | 10 | 9 | 8 | 7 | 6 | 5 | 4 | 3 | 2 | 1 | 0 |
|---|---|---|---|---|---|---|---|---|---|---|---|---|
| person number | 1 | 2 | 3 | 4 | 5 | 6 | 7 | 8 | 9 | 10 | 11 | 12 |

The number of handshakes is

$$11 + 10 + 9 + 8 + 7 + 6 + 5 + 4 + 3 + 2 + 1 = 66.$$

For 12 students there will be 66 handshakes.

## • PROBLEM 3-81

The letters below represent consecutive integers on a number line.

a   b   c   d   e   f   g   h   i   j

If $2b + f = 13$, what is the value of $b$.

## SOLUTION:

The correct response is:

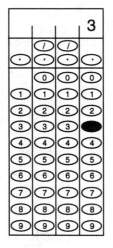

On a number line the numbers become greater as we move to the right.

Since the letters represent consecutive integers and $f$ is four spaces to the right of $b$, we can say that $b + 4 = f$

$$2b + f = 13$$
$$2b + (b + 4) = 13$$
$$2b + b + 4 = 13$$
$$3b + 4 = 13$$
$$3b = 9$$
$$b = 3$$

### • PROBLEM 3-82

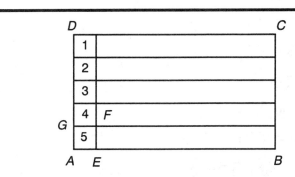

If *ABCD* is a rectangle and squares 1 through 5 each have an area of 4, and $\overline{EB} = 6\overline{AE}$, what is the area of *ABCD*?

## SOLUTION:

The correct response is:

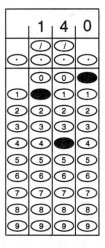

Since *AEFG* is a square with an area of 4, then each of its sides must equal 2. This is because the area of a square = $s^2$. Thus, $\overline{AG} = 2$.

Because $\overline{AG}$ is 2 and $\overline{AD}$ is 5 times as long as $\overline{AG}$,

$\overline{AD} = 5 \times 2 = 10$.

We also know that $\overline{EB} = 6\overline{AE}$. We know that $\overline{AE} = 2$, because all sides of a square are equal.

$\overline{EB} = 6\overline{AE}$

$\overline{EB} = 6 \times z$

$\overline{EB} = 12$

Also,

$$\overline{AE} + \overline{EB} = \overline{AB}$$

$$2 + 12 = \overline{AB}$$

$$14 = \overline{AB}$$

Thus, $AB = 14$ and $\overline{AD} = 10$.

$$\text{Area} = \text{length} \times \text{width}$$

$$\text{Area } ABCD = \overline{AB} \times \overline{AD}$$

$$= 14 \times 10$$

$$= 140$$

## • PROBLEM 3–83

If $6y + 3x + 8xy - 94 = 0$ and $x + 5 = 13$, then $y + 4 = ?$

## SOLUTION:

The correct response is:

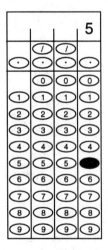

$$6y + 3x + 8xy - 94 = 0$$

$$x + 5 = 13$$

We can solve for $x$.

$$x + 5 = 13$$

$$x = 13 - 5$$

$$x = 8$$

We can now substitute this value for $x$ in the first equation.

$$6y + 3x + 8xy - 94 = 0$$

$$6y + 3(8) + 8(8)y - 94 = 0$$

$$6y + 24 + 64y - 94 = 0$$

$$70y - 70 = 0$$

$$70y = 70$$

$$y = \frac{70}{70}$$

$$y = 1$$

Therefore, $y + 4 = 1 + 4 = 5$.

### • PROBLEM 3-84

Two people are 33 miles apart. They begin to walk towards each other along a straight line at the same time. If one walks at the rate of 4 miles per hour while the other walks at the rate of 7 miles per hour, in how many hours will they meet?

## SOLUTION:

The correct response is:

Since this problem involves distance between two points, we use the distance formula.

Distance = rate × time

We know that the distance of the first walker plus the distance of the second walker is 33 miles.

Distance of 1st walker + Distance of 2nd walker = 33

We are given that the rate of the first walker is 4 miles/hr and that of the second walker is 7 miles/hr.

(rate) (time of 1st walker) + (rate) (time of 2nd walker) = 33

4 (time of 1st walker) + 7(time of 2nd walker) = 33

We know that both walkers walked for the same length of time, which we will label $T$.

Thus,

$$4T + 7T = 33$$
$$11T = 33$$
$$T = 3 \text{ hours}$$

They each walked for 3 hours.

## • PROBLEM 3–85

What is the largest integer for which $(a + 3)(a - 7)$ is negative?

## SOLUTION:

The correct response is:

$(a + 3) (a - 7)$ is negative

For the product of $(a + 3) (a - 7)$ to be negative, one of the terms must be positive and the other must be negative.

In looking at the terms $(a + 3)$ and $(a - 7)$, we can see that $(a + 3)$ will always be greater than $(a - 7)$.

Because $(a + 3)$ is a greater term, we want to find out the range of the values for which $(a + 3)$ will be positive. That is, $(a + 3) > 0$.

$a + 3 > 0$

$a + 3 - 3 > 0 - 3$

$a > - 3$

Thus, if $a > - 3$, then $(a - 3)$ will be positive.

Now we must determine when $(a - 7)$ will be negative,

$a - 7 < 0$

$a - 7 + 7 < 0 + 7$

$a < 7$

Thus, the product of $(a - 7)$ and $(a + 3)$ will be negative when $a < 7$ and $a > - 3$.

$- 3 < a < 7$

Therefore, the largest integer that will yield a negative number for $(a - 7)$ and a positive number for $(a + 3)$ would be 6.

### • PROBLEM 3-86

A student reaching into a refrigerator realizes that the solid container on the top shelf has a volume of $4/5\pi^2x^8$. Before closing the refrigerator, the student also notices that the solid container on the bottom shelf of the refrigerator has a volume of $5/4\pi^2x^8$. What percent of the container on the bottom shelf is the container on the top shelf?

## SOLUTION:

The correct response is:

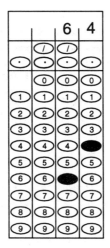

$$\frac{\text{container on top shelf}}{\text{container on bottom shelf}} = \frac{\frac{4}{5}\pi^2x^8}{\frac{5}{4}\pi^2x^8}$$

$$= \frac{\frac{4}{5}}{\frac{5}{4}}$$

$$= \frac{4}{5} \times \frac{4}{5} = \frac{16}{25}$$

$$= .64$$

$$= 64\%$$

## • PROBLEM 3-87

A coach informed his team that the average score for players on the team was *m*. The coach further said that the average score for the first 15 players was 70. There were 25 players on the team, and the range of the scores was from 60 to 100, inclusive. What is the greatest possible difference between the highest and lowest possible values of *m*?

### SOLUTION:

The correct response is:

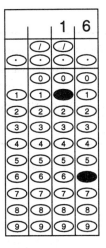

The first 15 players had an average score of 70. These 15 players scored exactly $70 \times 15 = 1,050$ points.

Since there were 25 players in total on the team, there were $25 - 15 = 10$ remaining players. To find the lowest possible value of *m*, which is the lowest possible average, we must assume that each of the 10 remaining players produced the lowest possible score.

The lowest possible score was 60. If each of the remaining 10 players scored 60, these 10 players would have scored $(60)(10) = 600$ points.

The total low score for all 25 players is $1,050 + 600 = 1,650$ points.

1,650 points scored by 25 people would be an average of $1,650 \div 25$ or 66 points per player.

To find the highest possible value of *m*, which is the highest possible average, we must assume that the 10 remaining players produced the highest score possible.

The highest possible score was 100. If each of the remaining 10 players scored 100, these 10 players would have scored $10 \times 100 = 1,000$ points.

The total high score for all 25 players would be $1,050 + 1,000$ or 2,050 points.

2,050 points scored by 25 players would average $2,050 \div 25$ or 82 points per team member.

Thus, the highest possible average would be 82 and the lowest would be 66. Thus, the difference is $82 - 66 = 16$ points.

## • PROBLEM 3-88

A child has six tiles. Each tile is emblazoned with one of the following letters: G, H, I, J, K, or L. How many different three-letter arrangements (such as HKG) can the child create if G is always the last letter in the arrangement and one of the letters in the arrangement is K?

## SOLUTION:

The correct response is:

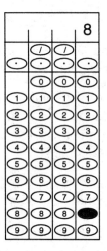

We note that the order of the letters is important in the problem since *G* must always be last and *K* must be one of the other two letters.

We can first write all the arrangements where *K* is the first letter and *G* is the last letter.

{*KHG, KIG, KJG, KLG*} 4 arrangements

Then we write all the arrangements where *K* is the second letter and another letter, other than *K* or *G,* is the first letter.

{*HKG, IKG, JKG, LKG*} 4 arrangements

This gives a total of 8 arrangements where *G* is always the last letter and one of the letters in the arrangement is *K.*

## • PROBLEM 3-89

What is the smallest positive integer that is divisible by each of the following numbers: 3, 4, 5, and 6?

## SOLUTION:

The correct response is:

We are being asked to find the least common multiple. To do this take the highest number, 6, and then ask whether 3, 4, and 5 goes evenly into multiples of 6. The lowest number for which this occurs will be the answer.

We see that 60 is divisible by 5, 4, 3.

Therefore, 60 is the smallest positive integer that is divisible by 3, 4, 5, and 6.

## • **PROBLEM 3-90**

If, in the equations below, $h = 6$, then what is the value of $g$?

$2g + 4h + 6i = 22$

$2h + 2i = 10$

### SOLUTION:

The correct response is:

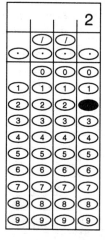

$$2g + 4h + 6i = 22 \qquad (1)$$

$$2h + 2i = 10 \qquad (2)$$

$$h = 6$$

Substitute $h = 6$ in equation (2)

$$2(6) + 2i = 10$$

$$12 + 2i = 10$$

$$2i = -2$$

$$i = -1$$

We know now that $h = 6$ and $i = -1$. We can substitute these values into equation (1).

$$2g + 4h + 6i = 22$$

$$2g + 4(6) + 6(-1) = 22$$

$$2g + 24 - 6 = 22$$

$$2g + 18 = 22$$

$$2g = 4$$

$$g = 2$$

## • PROBLEM 3-91

If $m + 2m + 3m = 20 + m$, then what does $m = $ ?

## SOLUTION:

The correct response is:

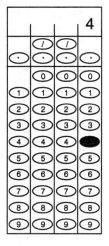

$$m + 2m + 3m = 20 + m$$

Combining like terms, we get

$$6m = 20 + m$$

$$6m - m = 20$$

$$5m = 20$$

$$m = \frac{20}{5}$$

$$m = 4$$

## • PROBLEM 3–92

$$35M$$
$$+ NM$$
$$M28$$

In the above correctly added addition problem, each letter stands for exactly one number. What is the value of *N*?

## SOLUTION:

The correct response is:

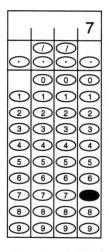

Reading down the column on the far right, we see

$M + M$ = a number which ends in 8.

What values of *M* are possible?

If $M = 4$, $M + M = 8$ (a number which ends in 8).

If $M = 9$, $M + M = 18$ (a number which ends in 8).

Thus, $M = 4$ or 9 are the two possible values.

First let's try $M = 4$ to see if the given addition problem will work out correctly when we use this value. We substitute 4 for *M* every place where *M* appears. This gives

$$354$$
$$+ N4$$
$$428$$

What values of *N* are possible when $M = 4$?

Reading down the middle column, we see

$5 + N$ = a number which ends in 2.

(We did *not* need to carry any value over from the column on he far right, since our choice $M = 4$ implies $M + M = 8$. In the case where $M = 9$, and $M + M = 18$, we would need to carry a 1.)

Trying different values for $N$ in the expression $5 + N$, we find $N = 7$ is the only value for $N$ which results in a number ending in 2. The next step is to substitute 7 for $N$, in order to see if the addition works for these values.

$$
\begin{array}{r}
354 \\
+\ 74 \\
\hline
428
\end{array}
$$

The addition is correct, so $M = 4$ and $N = 7$ are correct. If you were to go back to the beginning and repeat the same process using $M = 9$, you would not get a correct sum.

The answer to this problem is thus $N = 7$.

## • PROBLEM 3-93

A child draws three triangles in the sand and labels them triangle *A*, triangle *B*, and triangle *C*. Triangle *A* is three times the area of triangle *B*, triangle *B* is three times the area of triangle *C*, and triangle *B* has an area of 3. If the areas of all three triangles are added together, what would be their sum?

## SOLUTION:

The correct response is:

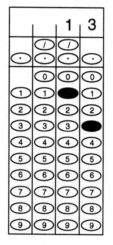

area $\Delta B = 3$

area $\Delta A = 3(\text{area } \Delta B) = 3 \times 3 = 9$

area $\Delta B = 3(\text{area } \Delta C)$

If *B* is 3 times *C*, to make *C* and *B* equal, we must divide *B* by 3. Therefore ,

$$\text{area } \Delta C = \frac{\text{area } \Delta B}{3} = \frac{3}{3}$$

area $\Delta C = 1$

Thus, the sum of the area equals

$1 + 3 + 9 = 13$.

## • PROBLEM 3-94

For all numbers $c$ and $d$, let $\lozenge$ be defined by the following equation

$$c \lozenge d = d^2 + cd - c.$$

If $3 \lozenge g = 25$, what is the positive value of $g$?

## SOLUTION:

The correct response is:

If $\lozenge$ defines the relationship of the values on either side, then

$$c \lozenge d = d^2 + cd - c$$

and $\quad 3 \lozenge g = g^2 + 3g - 3$

(meaning $c = 3$, $d = g$ by the definition)

$$= g^2 + 3g - 3$$

According to the statement of the problem

$$3 \lozenge g = g^2 + 3g - 3 = 25$$

(Solving this equation for $g$, set the right side equal to zero by adding $-25$ to both sides.)

$$g^2 + 3g - 28 = 0$$

(Factor the left side)

$$(g + 7)(g - 4) = 0$$

(Set each factor equal to zero)

$$g + 7 = 0 \qquad g - 4 = 0$$

$$g = -7 \qquad g = 4$$

Since the question asks for the positive value of $g$, $g = 4$.

## • PROBLEM 3-95

A list of numbers has been arranged such that each number in the list is 8 less than the number that precedes it. If 105 is the ninth number in the list, what is the fourth number in the list?

## SOLUTION:

The correct response is:

We need to set up a list of 9 numbers. The ninth number is 105. Each number on the list is 8 less than the number which precedes it. We can fill in the list as follows:

| 1st | 2nd | 3rd | 4th | 5th | 6th | 7th | 8th | 9th |
|-----|-----|-----|-----|-----|-----|-----|-----|-----|
| 159 | 141 | 153 | 145 | 137 | 129 | 121 | 113 | 105 |

The fourth number on the list is 145.

## • PROBLEM 3–96

Two identical humidifers vaporize water at exactly the same uniform rate. Each humidifier has been filled to its maximum capacity. It takes exactly 8 hours for each humidifier to completely vaporize a tank of water. The first humidifier is turned on at exactly 6:00 P.M. The second humidifier is turned on exactly one hour later. How many hours will it take before the water remaining in the second tank is twice the water remaining in the first tank?

## SOLUTION:

The correct response is:

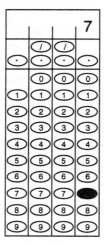

We can consider the following chart. Since we are told that the tanks require 8 hours to completely vaporize, and because they vaporize at a uniform rate, we may assume that each tank holds 8 gallons of water, and therefore, each humidifier vaporizes 1 gal of water each hour.

| Before humidifier is turned on | First Tank–8 gal. | Second Tank–8 gal. |
|---|---|---|
| End of 1 hr | 7 gal | 8 gal |
| End of 2 hrs | 6 gal | 7 gal |
| End of 3 hrs | 5 gal | 6 gal |
| End of 4 hrs | 4 gal | 5 gal |
| End of 5 hrs | 3 gal | 4 gal |
| End of 6 hrs | 2 gal | 3 gal |
| End of 7 hrs | 1 gal | 2 gal |

After 7 hours there are twice as many gallons in the second tank as in the first tank.

## • PROBLEM 3-97

The price of a compact disc player was reduced from $200.00 to $147.59. What was the percentage decrease in the price of the unit?

## SOLUTION:

The correct response is:

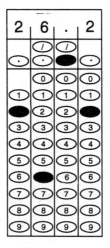

The price of the compact disc player was reduced from $200 to $147.59. The amount of decrease is

$$\$200.00 - \$147.59 = \$52.41$$

The percentage decrease is the

$$\frac{\text{amount of the decrease}}{\text{original amount of unit}}$$

$$\frac{52.41}{200.00} = .26205$$

To obtain a percent from a fraction we move the decimal point two places to the right.

$$.26205 = 26.205\%$$

## • PROBLEM 3-98

In triangle *ABC*, $\overline{AB}$ = 6 and $\overline{AC}$ = 9.2. If $\overline{BC}$ is an integer, what is the smallest possible perimeter of the triangle?

## SOLUTION:

The correct response is:

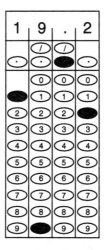

In order to find the smallest possible perimeter, we must find the smallest possible value for side $\overline{BC}$.

When any two sides of a triangle are added together, the sum of these two sides must be greater than the third side.

Thus,

$$\overline{BC} + \overline{AB} > \overline{AC}$$

$$\overline{BC} + 6 > 9.2$$

$$\overline{BC} > 9.2 - 6$$

$$\overline{BC} > 3.2$$

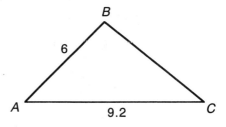

We are told that $\overline{BC}$ is an integer. The smallest integer greater than 3.2 is 4. Thus, the smallest possible value of $\overline{BC}$ is 4.

The perimeter of the triangle is therefore

$$\overline{AB} + \overline{AC} + \overline{BC} = 6 + 9.2 + 4$$

$$= 19.2$$

## • PROBLEM 3-99

A bus driving at an average speed of 55 miles per hour takes the basketball team to a game in 4 hours. Because of snowy weather conditions, while following the same route home, the bus must average 40 miles per hour. How many minutes more will the return trip take than the original trip?

## SOLUTION:

The correct response is:

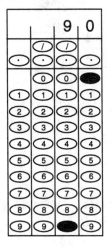

Since the bus averages 55 miles per hour for 4 hours, the length of the trip is found using the distance formula.

Distance = rate × time

= 55 × 4 = 220 miles

Because the bus followed the same route, the return trip must also be 220 miles. We can now substitute into the distance formula to compute the time.

$$D = R \times T$$

$$220 = 40 \times T$$

$$220 = 40T$$

$$\frac{220}{40} = T$$

$$5.5 = T$$

Therefore, the return trip took 5.5 hours. The return trip was 1.5 hrs longer. $(5.5 - 4.0 = 1.5)$

Finally, we can convert 1.5 hours into minutes by multiplying by the number of minutes in an hour.

$60 \times 1.5 = 90$

## • PROBLEM 3-100

If a teacher travels from point $M$ to point $O$, by first passing through point $N$, the teacher will have traveled a distance of 225 miles. If the rate of travel while going from point $M$ to point $N$ is 15 miles per hour and the rate of travel while going from point $N$ to point $O$ is 30 miles per hour, and the trip has a total travel time of 9 hours, what is the distance, in miles, from point $N$ to point $O$?

## SOLUTION:

The correct response is:

The distance from point $M$ to point $O$ plus the distance from point $N$ to point $O$ equals 225.

Distance $MN$ + Distance $NO$ = 225

rate × time (of $MN$) + rate × time ($NO$) = 225

15 × time (of $MN$) + 30 × time ($NO$) = 225

We are given that it took 9 hours for the trip.

Let time of $MN = x$

time of $NO = 9 - x$

Substituting these values in the above equation, we get

$$15 \times x + 30 \times (9 - x) = 225$$

$$15x + 270 - 30x = 225$$

$$270 - 225 - 15x = 0$$

$$45 = 15x$$

$$3 = x$$

$$6 = 9 - x$$

Hence, the time to cover distance $MN = 3$ hours and the time to cover distance $NO = 6$ hrs.

Distance $NO = $ (30 miles/hr) 6 hr

$$= 180 \text{ miles}$$

# Appendix

# REFERENCE TABLE

## SYMBOLS AND THEIR MEANINGS

| | | | |
|---|---|---|---|
| = | is equal to | | is less than or equal to |
| | is unequal to | ™ | is greater than or equal to |
| < | is less than | ‖ | is parallel to |
| > | is greater than | ⊥ | is perpendicular to |

## FORMULAS

| DESCRIPTION | FORMULA |
|---|---|
| **Area ($A$) of a:** | |
| square | $A = s^2$; where $s$ = side |
| rectangle | $A = lw$; where $l$ = length, $w$ = width |
| parallelogram | $A = bh$; where $b$ = base, $h$ = height |
| triangle | $A = \frac{1}{2} bh$; where $b$ = base, $h$ = height |
| circle | $A = \pi r^2$; where $\pi$ = 3.14, $r$ = radius |
| **Perimeter ($P$) of a:** | |
| square | $P = 4s$; where $s$ = side |
| rectangle | $P = 2l + 2w$; where $l$ = length, $w$ = width |
| triangle | $P = a + b + c$; where $a$, $b$, and $c$ are the sides |
| circumference (C) of a circle | $C = \pi d$; where $\pi$ = 3.14, $d$ = diameter = $2r$ |
| **Volume ($V$) of a:** | |
| cube | $V = s^3$; where $s$ = side |
| rectangular container | $V = lwh$; where $l$ = length, $w$ = width, $h$ = height |
| **Pythagorean Theorem** | $c^2 = a^2 + b^2$; where $c$ = hypotenuse, $a$ and $b$ are legs of a right triangle |
| **Distance ($d$):** | |
| between two points in a plane | $d = \sqrt{(x_2 - x_1)^2 + (y_2 - y_1)^2}$ where $(x_1, y_1)$ and $(x_2, y_2)$ are two points in a plane |
| as a function of rate and time | $d = rt$; where $r$ = rate, $t$ = time |
| **Mean** | mean = $\dfrac{x_1 + x_2 + \ldots + x_n}{n}$ where the $x$ s are the values for which a mean is desired, and $n$ = number of values in the series |
| **Median** | median = the point in an ordered set of numbers at which half of the numbers are above and half of the numbers are below this value |
| **Simple Interest ($i$)** | $i = prt$; where $p$ = principal, $r$ = rate, $t$ = time |
| **Total Cost ($c$)** | $c = nr$; where $n$ = number of units, $r$ = cost per unit |

# Index

Numbers on this page refer to <u>PROBLEM NUMBERS</u>, not page numbers.

# INDEX

Numbers on this page refer to <u>PROBLEM NUMBERS</u>, not page numbers.

## STUDENT-PRODUCED RESPONSE

### Arithmetic

### Algebra

### Geometry

**QUICK STUDY & REVIEW FOR THE NEW SAT I**

for students with limited study time and those who want a "refresher" before taking the test

**REA** *Research & Education Association*

*Available at your local bookstore or order directly from us by sending in coupon below.*

*Available at your local bookstore or order directly from us by sending in coupon below.*

# The High School Tutors®

  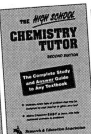   

     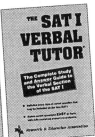

The **HIGH SCHOOL TUTOR** series is based on the same principle as the more comprehensive **PROBLEM SOLVERS**, but is specifically designed to meet the needs of high school students. REA has revised all the books in this series to include expanded review sections and new material. This makes the books even more effective in helping students to cope with these difficult high school subjects.

---

*If you would like more information about any of these books,*
*complete the coupon below and return it to us or go to your local bookstore.*

**RESEARCH & EDUCATION ASSOCIATION**
61 Ethel Road W. • Piscataway, New Jersey 08854
Phone: (908) 819-8880

**Please send me more information about your High School Tutor books.**

Name _____

Address _____

City _____ State _____ Zip _____

*Available at your local bookstore or order directly from us by sending in coupon below.*

# "The ESSENTIALS" of HISTORY

REA's **Essentials of History** series offers a new approach to the study of history that is different from what has been available previously. Compared with conventional history outlines, the **Essentials of History** offer far more detail, with fuller explanations and interpretations of historical events and developments. Compared with voluminous historical tomes and textbooks, the **Essentials of History** offer a far more concise, less ponderous overview of each of the periods they cover.

The **Essentials of History** provide quick access to needed information, and will serve as a handy reference source at all times. The **Essentials of History** are prepared with REA's customary concern for high professional quality and student needs.

## UNITED STATES HISTORY
**1500 to 1789** From Colony to Republic
**1789 to 1841** The Developing Nation
**1841 to 1877** Westward Expansion & the Civil War
**1877 to 1912** Industrialism, Foreign Expansion & the Progressive Era
**1912 to 1941** World War I, the Depression & the New Deal
**America since 1941:** Emergence as a World Power

## EUROPEAN HISTORY
**1450 to 1648** The Renaissance, Reformation & Wars of Religion
**1648 to 1789** Bourbon, Baroque & the Enlightenment
**1789 to 1848** Revolution & the New European Order
**1848 to 1914** Realism & Materialism
**1914 to 1935** World War I & Europe in Crisis
**Europe since 1935:** From World War II to the Demise of Communism

## WORLD HISTORY
Ancient History (4,500 BC to AD 500)
The Emergence of Western Civilization
Medieval History (AD 500 to 1450)
The Middle Ages

## CANADIAN HISTORY
**Pre-Colonization to 1867**
The Beginning of a Nation
**1867 to Present**
The Post-Confederate Nation

*If you would like more information about any of these books,*
*complete the coupon below and return it to us or go to your local bookstore.*

---

**RESEARCH & EDUCATION ASSOCIATION**
61 Ethel Road W. • Piscataway, New Jersey 08854
Phone: (908) 819-8880

**Please send me more information about your History Essentials Books**

Name _____

Address _____

City _____ State _____ Zip _____

# REA's **Problem Solvers**

The "PROBLEM SOLVERS" are comprehensive supplemental textbooks designed to save time in finding solutions to problems. Each "PROBLEM SOLVER" is the first of its kind ever produced in its field. It is the product of a massive effort to illustrate almost any imaginable problem in exceptional depth, detail, and clarity. Each problem is worked out in detail with a step-by-step solution, and the problems are arranged in order of complexity from elementary to advanced. Each book is fully indexed for locating problems rapidly.

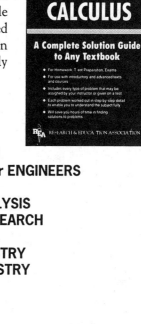

ACCOUNTING
ADVANCED CALCULUS
ALGEBRA & TRIGONOMETRY
AUTOMATIC CONTROL
   SYSTEMS/ROBOTICS
BIOLOGY
BUSINESS, ACCOUNTING, & FINANCE
CALCULUS
CHEMISTRY
COMPLEX VARIABLES
COMPUTER SCIENCE
DIFFERENTIAL EQUATIONS
ECONOMICS
ELECTRICAL MACHINES
ELECTRIC CIRCUITS
ELECTROMAGNETICS
ELECTRONIC COMMUNICATIONS
ELECTRONICS
FINITE & DISCRETE MATH
FLUID MECHANICS/DYNAMICS
GENETICS
GEOMETRY

HEAT TRANSFER
LINEAR ALGEBRA
MACHINE DESIGN
MATHEMATICS for ENGINEERS
MECHANICS
NUMERICAL ANALYSIS
OPERATIONS RESEARCH
OPTICS
ORGANIC CHEMISTRY
PHYSICAL CHEMISTRY
PHYSICS
PRE-CALCULUS
PROBABILITY
PSYCHOLOGY
STATISTICS
STRENGTH OF MATERIALS &
   MECHANICS OF SOLIDS
TECHNICAL DESIGN GRAPHICS
THERMODYNAMICS
TOPOLOGY
TRANSPORT PHENOMENA
VECTOR ANALYSIS

*If you would like more information about any of these books,*
*complete the coupon below and return it to us or visit your local bookstore.*

---

**RESEARCH & EDUCATION ASSOCIATION**
61 Ethel Road W. • Piscataway, New Jersey 08854
Phone: (908) 819-8880

**Please send me more information about your Problem Solver Books**

Name _____

Address _____

City _____ State _____ Zip _____

# MAXnotes®

## REA's Literature Study Guides

**MAXnotes®** are student-friendly. They offer a fresh look at masterpieces of literature, presented in a lively and interesting fashion. **MAXnotes®** offer the essentials of what you should know about the work, including outlines, explanations and discussions of the plot, character lists, analyses, and historical context. **MAXnotes®** are designed to help you think independently about literary works by raising various issues and thought-provoking ideas and questions. Written by literary experts who currently teach the subject, **MAXnotes®** enhance your understanding and enjoyment of the work.

Available **MAXnotes®** include the following:

| | | |
|---|---|---|
| Absalom, Absalom! | Heart of Darkness | Of Mice and Men |
| The Aeneid of Virgil | Henry IV, Part I | On the Road |
| Animal Farm | Henry V | Othello |
| Antony and Cleopatra | The House on Mango Street | Paradise Lost |
| As I Lay Dying | Huckleberry Finn | A Passage to India |
| As You Like It | I Know Why the Caged | Plato's Republic |
| The Autobiography of | Bird Sings | Portrait of a Lady |
| Malcolm X | The Iliad | A Portrait of the Artist |
| The Awakening | Invisible Man | as a Young Man |
| Beloved | Jane Eyre | Pride and Prejudice |
| Beowulf | Jazz | A Raisin in the Sun |
| Billy Budd | The Joy Luck Club | Richard II |
| The Bluest Eye, A Novel | Jude the Obscure | Romeo and Juliet |
| Brave New World | Julius Caesar | The Scarlet Letter |
| The Canterbury Tales | King Lear | Sir Gawain and the |
| The Catcher in the Rye | Les Misérables | Green Knight |
| The Color Purple | Lord of the Flies | Slaughterhouse-Five |
| The Crucible | Macbeth | Song of Solomon |
| Death in Venice | The Merchant of Venice | The Sound and the Fury |
| Death of a Salesman | The Metamorphoses of Ovid | The Stranger |
| The Divine Comedy I: Inferno | The Metamorphosis | The Sun Also Rises |
| Dubliners | Middlemarch | A Tale of Two Cities |
| Emma | A Midsummer Night's Dream | The Taming of the Shrew |
| Euripides' Medea & Electra | Moby-Dick | The Tempest |
| Frankenstein | Moll Flanders | Tess of the D'Urbervilles |
| Gone with the Wind | Mrs. Dalloway | Their Eyes Were Watching God |
| The Grapes of Wrath | Much Ado About Nothing | To Kill a Mockingbird |
| Great Expectations | My Antonia | To the Lighthouse |
| The Great Gatsby | Native Son | Twelfth Night |
| Gulliver's Travels | 1984 | Uncle Tom's Cabin |
| Hamlet | The Odyssey | Waiting for Godot |
| Hard Times | Oedipus Trilogy | Wuthering Heights |